Green Building Costs

Sustainability has become a driver of innovation in the built environment, but the affordability of sustainable building remains a significant challenge. This book takes a critical view of the real construction cost of green building. It provides readers with a non-biased evaluation based on empirical construction cost data and sheds light on the affordability of sustainable buildings.

Chapters are presented in three parts. The first part lays the foundation to demystify the perception of green buildings being expensive to construct by providing empirical evidence that green buildings, even net-zero buildings, are not necessarily more expensive to build than conventional buildings. The second part presents empirical evidence, common misperceptions of a higher green building construction cost are debunked. The author offers a new framework to explain the construction cost drivers and differences of sustainable buildings: the project characteristics and project team characteristics (human factors). The third part directs the readers' attention to the important role that human factors play in controlling and reducing construction costs, with a focus on the project design team. A lack of skills, expertise, and experience during the design phase is likely to be the biggest contributor to higher construction costs. Empirical analysis, case studies on LEED-certified buildings, and interviews with project teams are used to present a pathway to more affordable green building at the end.

This will be a crucial resource for students and professionals in architecture, engineering, construction management, and planning and energy policy.

Ming Hu is an Associate Professor at School of Architecture and Associate Professor at College of Engineering, University of Notre Dame, USA. Her research activities center on how to decarbonize the built environment through net-zero impact and healthy building design, and to understand how the (smart) technologies might be employed to reduce the impact from built environment to ecosystem.

Green Building Costs
The Affordability of Sustainable Design

Ming Hu

Routledge
Taylor & Francis Group

LONDON AND NEW YORK

Designed cover image: Ming Hu

First published 2024
by Routledge
4 Park Square, Milton Park, Abingdon, Oxon OX14 4RN

and by Routledge
605 Third Avenue, New York, NY 10158

Routledge is an imprint of the Taylor & Francis Group, an informa business

© 2024 Ming Hu

British Library Cataloguing-in-Publication Data
A catalogue record for this book is available from the British Library

Library of Congress Cataloging-in-Publication Data
Names: Hu, Ming, 1975– author.
Title: Green building costs: the affordability of sustainable design/Ming Hu.
Description: Abingdon, Oxon; New York, NY: Routledge, 2024. |
Includes bibliographical references and index. |
Identifiers: LCCN 2023005084 | ISBN 9781032328096 (hbk) |
ISBN 9781032328089 (pbk) | ISBN 9781003316855 (ebk)
Subjects: LCSH: Sustainable buildings—Design and construction—Costs. |
Sustainable buildings—Costs.
Classification: LCC TH880 .H794 2024 |
DDC 690.028/6—dc23/eng/20230214
LC record available at https://lccn.loc.gov/2023005084

ISBN: 978-1-032-32809-6 (hbk)
ISBN: 978-1-032-32808-9 (pbk)
ISBN: 978-1-003-31685-5 (ebk)

DOI: 10.4324/9781003316855

Typeset in Univers
by codeMantra

Contents

Figures

Acknowledgments

I owe a debt of gratitude to the many people who supported this book and helped me in a number of ways.

I am indebted to Professor Emeritus Norman Crowe from the University of Notre Dame and Dr. Alexandra Staub from Pennsylvania State University for reviewing the draft and providing the feedback and advice. I would also like to thank my dissertation advisor, Dr. Miroslaw Skibniewski from the University of Maryland. He generously offered his mentorship, support, and guidance during my doctoral study.

I am also grateful to Fran Ford and Lydia Kessell, publisher and commissioning editor at Routledge/Taylor & Francis Group, who accepted my book proposal, and Jake Millicheap, editorial assistant at Routledge/Taylor & Francis Group, who shepherded the writing and publication phase. In particular, I would like to acknowledge Janna Christie for her professional and excellent editing help and support.

Finally, and most importantly, I would like to thank my husband Kai Hu, thank you for your unconditional support and love.

Part I

Looking back and perceptions

1 Introduction

ABSTRACT

Chapter 1 introduces the author's motivation for writing this book, the background and scope of the book, and overarching questions addressed by the author. It also outlines a proposed conceptual model to explain the green building construction cost. Finally, the organization of the book is detailed to facilitate readers' understanding.

1.1 MOTIVATION BEHIND THIS BOOK

1.1.1 Practice and research experience

The author's interest in writing a book about the affordability of sustainable building is rooted in her professional working experience and academic research expertise in sustainable building design and construction. The author has in-depth knowledge of sustainable rating systems and years of experience working as a project architect on many Leadership in Energy and Environmental Design (LEED)-certified sustainable buildings. An example is the Abu Dhabi National Oil Company headquarters, which has 75 stories and is LEED Gold certified. The total construction cost was around $538 million with a unit cost of $3,046/m^2 ($283/ft^2) (BESIX, n.d.), which is considerably lower than that of LEED buildings in the United States. According to a Cummings Construction market analysis, the average cost of a US high-rise office building (non-LEED) was around $7,104/m^2 ($660/ft^2) in 2021. Besides the lower labor costs in Abu Dhabi, other factors drove down the cost of this project, such as having a skilled and experienced integrated design team.

From a practitioner's perspective, over time, the author has observed the positive and negative impacts that design teams have on project costs and schedules. Those negative impacts are often unseen from public eyes. For example, some practitioners started a project with an assumed higher sustainable building construction cost and consequently disregard cost control; in other cases, the design teams specified unnecessary, expensive materials and products that do not directly contribute to the building's sustainability but instead promote the perception that sustainable building is expensive. While these compounding causes directly contribute to a higher construction cost for sustainable buildings, they are often ignored both in research and practice since the costs related

Figure 1.1
The National
Renewable
Energy Laboratory
Research Support
Facility by Dennis
Schroeder is
licensed under CC
BY-ND 2.0

to design decision-making represent less than 10%. Consequently, the project design team's impact is less studied and understood.

From a research aspect, the author's motivation to dive into this topic was a journal paper "A pathway for net-zero energy buildings: creating a case for zero cost increase" published by researchers from the National Renewable Energy Laboratory (NREL). NREL built the Research Support Facility (RSF) in Colorado, which is a 442,907-ft^2 (41,147 m^2) LEED Platinum building. It is also the first large-scale net-zero energy building in the United States (refer to Figure 1.1).

This building was completed in 2008 (phase 1) and 2010 (phase 2) with actual construction costs ($2,787/m^2 and $2,637/m^2, respectively) that were a third lower than those of a typical commercial construction of similar size and type in the United States (Torcellini et al., 2015). Two successful cost-control strategies were identified from this project: the *competitive procurement process* and *the integrated decision-making process* in the early design stages. The client (NREL) incorporated measurable performance goals into the project request for the proposal; therefore, when the team bid on this project, they treated the listed sustainability goals as an indispensable requirement within the allowed budget. After selecting the project team (included contractors and designers), NREL also integrated trade partners into the overall decision-making process in the early design stages to ensure that construction considerations were properly weighed during the design. Moreover, all the stakeholders agreed on the design decisions and related construction cost implications. Such an integrated approach is particularly useful for green building, where certain sustainable practices and technologies may not be familiar to all team members. Those unfamiliar technologies may contribute to a cost increase because risk and uncertainty are usually associated with a cost increase (Amiril et al., 2017). However, in the RSF project, risk and uncertainty were managed well by the integrated and capable design team, so the final construction cost of this first large-scale net-zero energy building was cheaper than a conventional building of the same scale.

Overall, a combination of firsthand experience and research findings motivated author to further explore and understand the cost of constructing a green building.

1.1.2 Market trends

There are multiple factors driving up the market demand for sustainable buildings. The first driver is the carbon emissions reduction goal. Building green has been recognized as one of the most effective strategies for overall energy consumption and carbon emissions reductions. As stated in the *World Green Building Trends 2018 SmartMarket Report*, organizations across the construction industry continue to shift toward more sustainable buildings and products (World Green Building Council, n.d.).

The second driver is real estate interest. A recently published article in the *New York Times*, titled "As Risks of Climate Change Rise, Investors Seek Greener Buildings," reported that mutual funds and exchange-traded funds invested nearly $300 billion in sustainable assets globally in 2020, nearly double that of the previous year. As stated by the chief investment officer of a Dutch investment firm, "Today, you don't sacrifice returns for sustainability, you create returns with sustainability." The effects of climate change are indeed altering the strategies of investment trends. If older buildings do not lower their carbon footprint, they are likely to suffer from depreciation of their assets as early as 2026 (Sisson, 2021).

The third driver is legislation incentives. For example, the "Better Buildings" initiative by the U.S. Department of Energy (DOE) is designed to prioritize investments in affordable, safe, and efficient homes and businesses powered by reliable, clean energy. All these drivers will boost the demand for sustainable buildings.

1.1.3 Overarching goals of this book

While it is promising that the market will eventually catch up with the potential value of green buildings, a profit-driven investment approach not only leads to higher prices for green buildings but also potential greenwashing. In addition, it is worrisome that the pursuit of an investor- and market-driven approach may boost the perception of sustainable buildings as being expensive. **The perceived high cost and market pursuit may further advance the reputation of sustainable building as an elite pursuit, eventually resulting in environmental injustice through pushing green building prices out of reach for low-income families and even middle-class families**. To this extent, this book has three overarching goals:

- Deliver the notion that sustainable building for all is within reach with appropriate approach via laying the foundation to debunk the expensive perception of sustainable building with actual construction cost data.

- Propose a framework for understanding the project design team's characteristics as influential cost drivers.
- Outline a path toward affordable green buildings by explaining the human factor effect and providing examples.

1.2 DESCRIPTION OF THE KNOWLEDGE GAP

1.2.1 Contradictions between drivers and perception

Regarding the drivers of sustainable building, Ramboll's *Sustainable Buildings Market Study 2019* showed the most important trends driving sustainable building activity are "Life cycle thinking and management" (71%), "Health and wellbeing" (58%), and "Increased focus on carbon neutrality" (53%) (Ramboll, 2019). These top drivers are indeed **different from** those of conventional buildings, such as program needs and investment gains. Such differences can impact the project financial planning and cost-benefit analysis. However, the Ramboll report did not provide a detailed explanation of the differences or through which mechanisms these drivers shift the sustainable building construction cost (SBCC) up or down. Both life cycle thinking and carbon neutrality call for low embodied carbon materials and building construction methods, which indicate local materials or less energy-intensive methods. Logically, these drivers lead to using more locally supplied materials that are affordable and familiar to local builders, consequently reducing the construction cost. This assumption contradicts the common perception of sustainable building being expensive.

1.2.2 Lack of studies on construction cost

To date, a large amount of research has focused on the benefits of sustainable building for users, clients, and society (Liu et al., 2014), such as incremental economic benefits through saving energy and improving the environment (Eichholtz et al., 2010). Gabay et al. (2014) pointed out how green labels affect the market rents and values of commercial spaces, potentially leading to a high resale value of a building. However, only a small portfolio of studies has investigated the actual construction cost (Zhang & Fuh, 1998). Regarding costs related to public sustainable building, there are limited literature and reports due to the difficulty of accessing actual construction cost data. Meanwhile, despite the widespread perception of sustainable building being expensive, the existing empirical studies and evidence to support this claim hold much ambiguity (refer to Chapter 2). Without a deeper understanding of the cost factors and drivers, creating effective strategies to control or reduce the construction cost is impossible. The ineffectiveness of certain current policies reflects a lack of supporting empirical evidence. Although much anecdotal evidence has been collected based on case studies, these non-systematic data and analyses do not generate enough importance to direct people's attention to this important topic.

1.2.3 Lack of consistency in construction cost data collection

As stated by Chegut et al. (2019), access to a comprehensive and consistent set of data on building construction costs is remarkably hard to obtain. Therefore, a variety of data collection methods have been employed. The *top* method for acquiring data is contacting the project team and collecting construction documents (Meron & Meir, 2017); occasionally, the researchers contact the building operators or owners for information as well (Hu & Skibniewski, 2021). Such methods require a collaborative working relation between practitioners and researchers and are time-consuming as researchers must organize and comb through the data.

The *second* most used data collection method is distributing questionnaires or surveys, sometimes followed by one-on-one interviews. This method is the most practical and convenient, but it produces much variation due to a certain subjectivity. Some early studies conducted by industry experts and green consultants showed no or minimal SBCC surcharge without explaining the data sources or the questions asked. Therefore, it was not clear whether the surveys focused on the practitioners' perception of green building costs or actual green building construction. In addition, due to the nature of the survey questions, it was difficult to gain a deeper understanding of which building system and components were most influential for the SBCC surcharge and why.

Use of a cost estimator is the *third* commonly used method for data collection. Some of the most influential SBCC studies were produced by cost estimators. For example, the Davis Langdon (later merged with AECOM in 2011) database was used for four different studies for projects in the United States, Australia, and China (Hong Kong). Davis Langdon has provided cost estimation and project management worldwide for many high-profile projects, such as Tate Modern and the Eden Project in the UK, the Abu Dhabi International Airport, the Clem Jones Tunnel (CLEM7) in Brisbane, and the former Transbay Terminal in San Francisco. The company's expertise, global practice coverage, and interest in practicing sustainable design could explain why it has conducted several studies and reported the construction cost of sustainable building to be comparable to that of conventional building. Davis Langdon's studies are among the first reports that question whether sustainable building is more expensive. However, since cost estimator databases are not publicly accessible, and its estimation method is generally not described or explained, the reliability and robustness of its study cannot be verified. Nonetheless, in the future, it will be beneficial to compare studies using data from different cost estimators' databases.

The *fourth* data collection method is relying on a publicly available database. This method presents transparency. Most publicly available databases, especially those sponsored by the government, such as the U.S. Census Bureau and Israel's Central Bureau of Statistics, provide detailed information about how the data are collected and analyzed as well as metadata files (Dwaikat & Ali, 2016). These public data are more transparent; furthermore, the same data collecting method can be replicated by other researchers or industry experts, which contributes tremendously to the consistency in construction cost studies. But level of details and accuracy is the shortcoming of some publicly available data. For instance, the U.S. Census Bureau uses the normalized building's sale price as a proxy for the construction cost. On its website, the U.S. Census Bureau's Survey of Construction (SOC) indicates that the total cost of a private, new single-family home is obtained by multiplying the number of units by an average construction cost per unit that is derived from the sale price. Further, there is a difference in determining the cost of an owner-built home and the cost of a developer-built home. For units built by a developer (to be sold or rented), the average construction cost is the average sales price at the time of start multiplied by the factor 0.8424. This factor eliminates the cost estimate of "nonconstruc-tion" items, such as raw land, marketing costs, closing costs, and movable appliances (US Census Bureau Construction Spending Survey, n.d.). Therefore, the construction cost data of a developer-built American home are obtained from the sales price using a single factor. This generalization omits differences in geography, building type, and construction quality as well as other cost factors. Although the data are publicly available and transparent, when researchers and practitioners used the database, the limitations need to be acknowledged, and the analysis report should be examined and justified.

Overall, the first method, where the data are acquired from project teams, developers, and building owners, is more reliable and accurate than the other methods. For example, Shrestha and Pushpala (2012) worked with Clark county public school districts in Nevada to gain access to the actual construction costs of green and non-green school buildings. Altogether, 30 green school build-ings and 30 non-green school buildings were studied, with the results showing the highest green building construction cost surcharge (46%) among all stud-ies. Although the parameters that influenced the cost were not defined, the researchers suggested including a cost and construction control in future school project planning to control the high green cost surcharge (Dwaikat & Ali, 2016).

1.2.4 Differing definitions of sustainability

The current terminology that describes sustainable building rating systems, cer-tificates, assessments, and related tools is inconsistent and unclear. A large body of research has examined the differences between rating systems. For example, Haapio and Viitaniemi (2008) reviewed 16 different building envi-ronmental assessment systems, including LEED, Athena, Building Research Establishment Environmental Assessment Method (BREEAM), Envest, BEAT 2002, Eco-Quantum, and Annex 31. The key issues identified were the under-standing of the tools and how these tools have affected decision-making in building green. Gou and Lau conducted a study on differences in green building certification systems at international, national, and local levels by choosing three green rating systems: HK-BEAM, China Green Building Label, and LEED V4. They explained the differences between the rating systems by using the conceptual-ism theory, which can be traced back to the fundamental divergence in lifestyles, preferences, and urban morphology, besides climatic variations (Gou & Lau, 2014). Nguyen and Altan compared five sustainable rating systems: BREEAM (UK), LEED (US), Comprehensive Assessment System for Built Environment Efficiency (CASBEE, Japan), GREEN STAR (Australia), and HK-BEAM. The results indicate that BREEAM and LEED scored the highest based on nine criteria (Nguyen et al., 2012).

There were two findings from the large body of studies comparing various sustainable building rating systems. First, there are similarities among the rating systems: 70% of the existing comparative studies focus on four leading rating systems: BREEAM, LEED, CASBEE, and China's "Three Star," which were developed geographically in the United States, the United Kingdom, Japan, and China (Doan et al., 2017). Among those different rating systems, there are three common categories: *energy efficiency*, *material resources,* and *indoor environment* (Doan et al., 2017).

Second, there are significant differences between the rating systems, which may lead to differences in buildings' actual performance. For example, Asdrubali et al. (2015) compared LEED (US) and ITACA (Italy) for residential buildings using two buildings located in Italy. Their findings revealed that for materials, LEED analyzes the number of components made from recycled materials used in construction, while ITACA is based on the number of components that are recyclable at the end of a building's life cycle. For indoor environmental quality, LEED includes more criteria than ITACA, giving special weight to the environmental effect of the construction phase (Asdrubali et al., 2015). Consequently, higher constraints are placed on the construction phase, which can influence the construction cost, method, and schedule.

The costs of pursuing different sustainability ratings and credits have not been sufficiently addressed since the existing studies focus more on the rating systems' impact on building performance rather than on cost. While several studies focused on how the construction cost related to the sustainable building score and credits, most studies used either a simulated (estimated) building cost or secondary data from a literature review (Tatari & Kucukvar, 2011; AlAwam & Alshamrani, 2021). Very few studies used a large firsthand data set of actual building costs. Consequently, the collective understanding of how sustainability impacts the construction cost is limited. Overall, the lack of a consistent definition and measurement of a building's sustainability increases difficulties in understanding the construction cost implications.

1.3 PROJECT DESIGN TEAM'S INFLUENCE

To propose the framework for understanding the project design team's impact on construction cost, the author first needs to assess the cost factors and drivers of sustainable buildings by using empirical data; therefore, collecting actual construction cost data is a critical task and contribution to the empirical research. The detailed description of data collected for this book can be found in Chapter 4.

Additionally, since sustainability has been perceived as a main cost driver, using empirical data to analyze the relation between the level of sustainability and the total construction cost will provide building practitioners, policy makers, and the public with non-biased evidence of whether sustainable buildings are affordable. Before the author introduces the proposed framework, an understanding of the theoretical foundation of the proposed framework is presented below.

1.3.1 Theoretical foundation of design decision's effect on construction cost

One of the most influential concepts regarding construction time and cost control is the effort curve, commonly known as the MacLeamy curve, which describes the relation between cost and change through the project stages. The effort curve was popularized by Patrick MacLeamy in 2004, but its origin was published by Professor Boyd C. Paulson Jr. in 1976 (Paulson, 1976). In a paper titled "Designing to Reduce Construction Costs," he first presented a diagram illustrating the level of influence on project costs (refer to Figure 1.2a). Paulson's first insights were that the decisions and commitments made during the early phases of a project (engineering and architectural design) have a significant impact on later expenditures. His second important observation was that efforts to suboptimize design costs by requiring competitive bidding for professional services were likely to produce much higher project costs in the long run (Paulson, 1976).

Since the publication of his concept paper in 1976, Paulson has become known for his involvement in large federally funded construction projects, such as the US urban rail projects BART, in Northern California, and Metrorail, in Washington, D.C. He also volunteered to oversee construction of Peninsula Habitat for Humanity's $2 million, 24-unit condominium project for low-income residents in California (Shwartz, 2005). Other researchers and practitioners tried to adopt and promote the project effort curve proposed by Paulson, but little attraction was gained until the early 2000s.

In 2004, Patrick MacLeamy introduced a curve, later known as the MacLeamy curve (refer to Figure 1.2b), in the Construction Users Roundtable in 2004 (CURT, 2004). While MacLeamy never acknowledged the resemblance of his curve to Paulson's project effort curve, the similarities between the two graphs are obvious. MacLeamy is an American architect and the chairman of buildingSMART International. MacLeamy has advanced the global implementation of building information modeling (BIM) to improve the quality and efficiency of the project delivery process (wiki). MacLeamy used this curve to present the following concept: Making optimized early design decisions can have a large impact on the project cost and schedule. The effort curve functions to illustrate the relation between cost and time as well as the importance of project planning and early decision-making.

Figure 1.2b demonstrates four important concepts about the relationship of design effort/cost (*Y*-axis) and the traditional phase of design and construction (*X*-axis). Each concept is represented by a colored line: (a) The red line (line 1)

Figure 1.2
(a) Project effort
curve (Paulson,
1976). (b)
MacLeamy curve
(Davies & Harty,
2013)

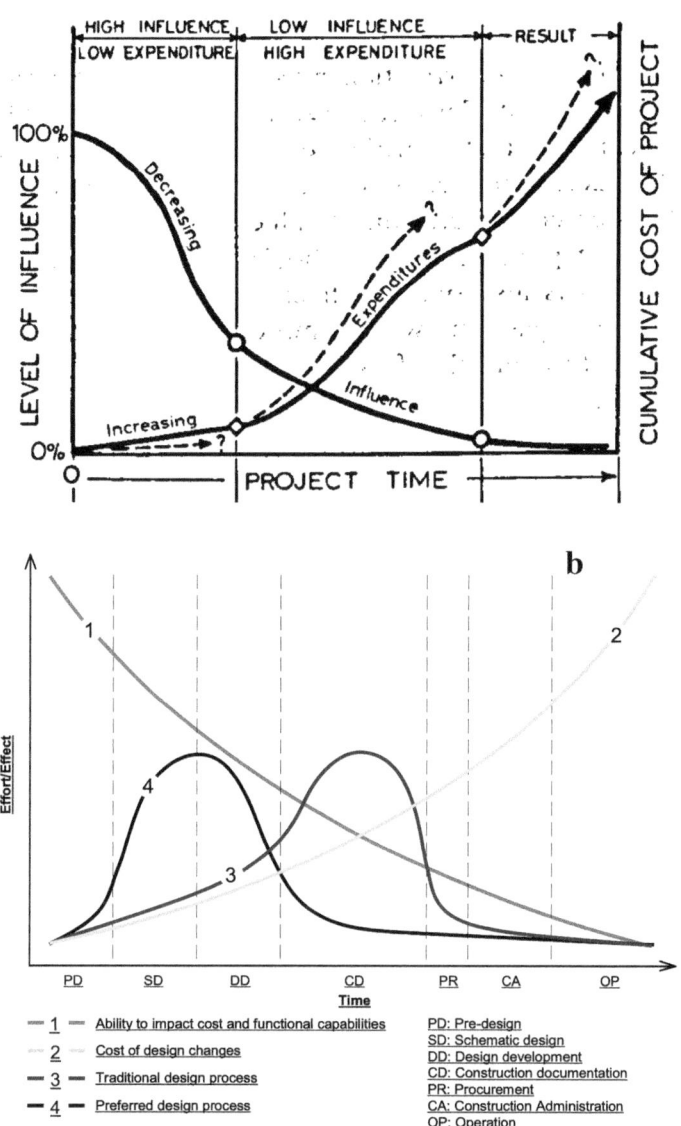

represents the team's decreasing ability to affect the project cost and other function capabilities as the project moves toward completion. (b) The green line (line 2) shows how the cost of design changes dramatically increases as the project progresses. (c) The blue line (line 3) represents the distribution of design efforts in a traditional building project, where the most effort (labor and cost) is allocated in the construction documentation (CD) stage – when the cost of change is high and the ability to control the cost is low. (d) The black line (line 4) suggests a new preferred design process, where design efforts are made in the early stages (PD-SD) – when the cost of change is low and the ability to control the cost is high.

The line 1 in the MacLeamy curve depicts the earliest possible decision-making for maximizing the ability to affect change and minimizing the potential cost of design changes (particularly those caused by large schedule delays and cost overruns). The author adopted this concept for this book to study the influence of design team characteristics and project characteristics of the early project stages (pre-design and schematic design) on the final construction cost. The design team characteristics are referred to as "project team characteristics" onward. The hypothesis indicates that early decisions have a large impact on the final construction cost; thus the project team characteristics' influence on the SBCC can be significant and impactful. Therefore, this book studies the *project team characteristics in the early design stages of sustainable projects*. The author focuses on two categorical variables: **project characteristics** and **project team characteristics**.

1.3.2 Proposed conceptual model
To guide readers through the book, a conceptual model has been created that hypothesizes the pattern of correlation and covariance among the set of constructs that impact the construction cost, as illustrated in Figure 1.3. The four main constructs discussed are project team characteristics, project characteristics, the level of sustainability, and the SBCC. Figure 1.3 also illustrates the hypothesized causal relations among these four constructs: a straight line with an arrow represents a causal relation, a curved line with an arrow represents covariance, a "+" sign represents a positive influence, and a "−" sign represents a negative influence. The following sections explain each construct.

1.3.3 Project characteristics
Project characteristics mainly focus on the project's physical and technical conditions. For project physical conditions, building type, construction type, and project scale are important influential factors (Sovacool et al., 2014) (Hwang & Low, 2012). In addition, the author also focuses on other important characteristics, such as technical complexity and supply chain maturity, which were speculated as influential factors to the conventional building construction cost from a conducted survey (Cheng, 2014). The exact definition and index used to measure the project characteristics are derived from the meta-analysis conducted by the author, which is explained in Chapter 3.

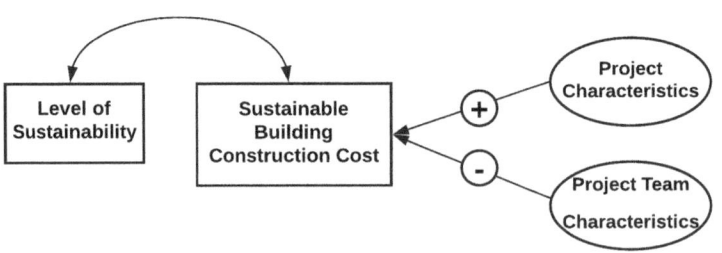

Figure 1.3
Proposed
conceptual model
and constructs

1.3.4 Project team characteristics

Several previous studies indicated that the skill and experience of the project manager and contractors are influential factors to the total construction cost (Cho et al., 2009). The potential interpretation is that the design team's skills and experience can mediate the negative effects of technical complexity on the construction cost of buildings. However, studies focusing on the design team's influence on sustainable building construction cost are very limited because these factors are more difficult to conceive due to their indirect and hidden nature (Cheng, 2014). Therefore, the author had to draw on her own practices and research experience, that is a similar design impact on the construction cost and a related explanation can be applied to sustainable building. For example, the heat pump has been recognized as one of the most efficient heating and cooling systems, and it is largely popular in European and Asian countries. Heat pumps constitute a third of the space heating sector in Japan. In Finland, in 2020, new single-family house heat pumps were installed in 75% of buildings (Hirvonen, 2021), while only 14% of American homes had electrical heat pumps according to US Residential Energy Consumption Survey data release in 2020, Table HC 6.1 (Energy Information Administrative, 2020). The lack of adoption and implementation of the heat pump as a sustainable technology is not caused by its technical complexity or hard costs – rather it is largely related to the design teams' and contractors' unfamiliarity of this technology in the United States (Hu et al., 2022). Therefore, the project team characteristics investigated in this project focus on the design team; specifically, the design team's *skill level, experience, motivation to innovate, communication, and collaboration*. The reason why these variables are used to present the project team characteristics are explained in Chapter 3.

1.3.5 Level of sustainability

In this book, the level of sustainability (LOS) is defined by the credits or level of certification received from the LEED rating system. LEED certification has four levels: Certified, Silver, Gold, and Platinum. To achieve each level, certain points are required, but the numbers of points required for the same level varies depending on the LEED rating system and which version is applied to the projects. For example, the current LEED v4.1 for Building Design and Construction (LEED BD+C) has a maximum of 110 points, and to achieve Certified, at least 40 points are required. Silver requires 50–59 points, Gold requires 60–79 points, and Platinum requires 80 or more points. However, the third version of LEED BD+C, released in late 2005, only has a total of 69 available points, where 26–32 points are required for Certified, 33–38 points for Silver, 39–51 points for Gold, and 52–69 points for Platinum. Furthermore, LEED v4.1 O+M, Existing Buildings Operations and Maintenance, has a total of 100 points, but the certification levels are the same as LEED BD+C. The projects included in this study are certified under different rating systems and/ or under different versions.

1.3.6 Construction cost

In this study, the author includes hard costs and soft costs in the construction cost. Hard costs are defined as a type of direct cost that can be traced back to the physical components or activities of the construction project, which includes the land cost, labor, materials, equipment, and the overhead cost from the contractor and developer (Ade & Rehm, 2020). In this study, the land cost and site cost are excluded. Soft costs are defined as the items not directly related to the physical construction of the buildings but that are necessary for the administration of a building project (Zahirah et al., 2013). Soft costs cover a wide range of cost items, from administration fees, design fees, and auditing fees (Ade & Rehm, 2020) to certification, commissioning, marketing, and taxes (Zahirah et al., 2013). Soft costs are mostly related to and affected by project team characteristics. Together, the hard costs and soft costs comprise the construction cost of a building.

1.4 ORGANIZATION OF BOOK

This book is composed of three parts. **Part I: Looking Back and Perceptions** (Chapters 1–3), **Part II: What Actual Data Tells Us** (Chapters 4 and 5), and **Part III: The Path Forward** (Chapters 6 and 7).

Part I attempts to demystify the perception of green building construction being expensive by providing existing empirical evidence. Chapter 1 introduces the motivation and focus of writing this book. Chapter 2 reviews existing literature about the SBCC to better understand the research activities and gaps regarding this topic in the last three decades. The knowledge gap on the SBCC is identified, and the SBCC surcharge is defined. Chapter 3 presents evidence from a meta-analysis conducted by the author on the global status of the SBCC surcharge, and differences per region, building type, and data sources are outlined. The results demonstrate a severe lack of empirical data for the actual SBCC, which leads to a large knowledge gap on why and how sustainable building is more expensive. Chapter 3 also indicates that the North American market, represented by sustainable projects in the United States, has the highest mean sustainable construction surcharge (SCS) and largest cost variance. Further investigations into the US market will provide in-depth knowledge on how and why sustainable building practices can be improved to create a more affordable and less risky sustainable building market.

To respond to this knowledge gap, in Part II, the author collected data from reliable sources to create an SBCC database specifically for this book, with a focus on the US market. In Chapter 4, the author explains the data source and collection procedure and provides descriptive statistics for the collected data. Chapter 5 centers on data analysis, with the author presenting the findings from the SBCC. After presenting empirical evidence, the author debunks common misperceptions of the green building construction cost being higher than its conventional counterpart. Lastly, the author offers a new framework to explain the construction cost drivers and differences of sustainable buildings: the project characteristics and project team characteristics.

Part III directs the readers' attention to the important role that project team characteristics play in controlling and reducing the construction cost. A path toward more affordable green building is presented based on the results from the case studies and interviews with project teams. Chapter 6 analyzes six built LEED projects, representing low-, median, and high-cost brackets, and focuses on the mechanism of how project team characteristics impact the total construction cost. Insights and further explanations are offered based on project team statements, online interviews, and other qualitative data from the case projects. Derived from the consensus built from the six case studies, in Chapter 7, the author debunks common misperceptions and offers explanations for the root causes of the misperception that sustainable building is expensive. Three key components for moving forward are presented. Then, Singapore's approach is presented as an example to help readers to see a pathway that is needed to make sustainable baseline and affordable. The proposed framework aims to help the public understand how to control the SBCC as well as instill confidence that affordable sustainable buildings are within reach.

REFERENCES

Ade, R., & Rehm, M. (2020). At what cost? An analysis of the green cost premium to achieve 6-homestar in New Zealand. *Journal of Green Building, 15*(2), 131–155. https://doi.org/10.3992/1943-4618.15.2.131

AlAwam, Y. S., & Alshamrani, O. S. (2021). Initial cost assessment stochastic model for green buildings based on LEED score. *Energy and Buildings, 245*, 111045. https://doi.org/10.1016/j.enbuild.2021.111045

Amiril, A., Nawawi, A., Takim, R., & Ab-Latif, S. N. F. (2017). The barriers to sustainable railway infrastructure projects in Malaysia. *The Social Sciences, 12*(5), 769–775.

Asdrubali, F., Baldinelli, G., Bianchi, F., & Sambuco, S. (2015). A comparison between environmental sustainability rating systems LEED and ITACA for residential buildings. *Building and Environment, 86*, 98–108.

BESIX. (n.d.). *ADNOC*. https://www.besix.com/en/projects/adnoc

Chegut, A., Eichholtz, P., & Kok, N. (2019). The price of innovation: An analysis of the marginal cost of green buildings. *Journal of Environmental Economics and Management, 98*, 102248.

Cheng, Y.-M. (2014). An exploration into cost-influencing factors on construction projects. *International Journal of Project Management, 32*(5), 850–860. https://doi.org/10.1016/j.ijproman.2013.10.003

Cho, K., Hong, T., & Hyun, C. (2009). Effect of project characteristics on project performance in construction projects based on structural equation model. *Expert Systems with Applications, 36*(7), 10461–10470. https://doi.org/10.1016/j.eswa.2009.01.032

Cummings Construction. (2021). *An in-depth look at construction costs per square foot in the United States*. https://ccorpinsights.com/costs-per-square-foot/

CURT. (2004). *Collaboration, integrated information, and the project lifecycle in building design, construction and operation* (WP-1202). https://kcuc.org/wp-content/uploads/2013/11/Collaboration-Integrated-Information-and-the-Project-Lifecycle.pdf

Davies, R., & Harty, C. (2013). Implementing 'Site BIM': A case study of ICT innovation on a large hospital project. *Automation in Construction, 30*, 15–24. https://doi.org/10.1016/j.autcon.2012.11.024.

Doan, D. T., Ghaffarianhoseini, G., & Naismith, N. (2017). A critical comparison of green building rating systems. *Building and Environment, 123*, 243–260. https://doi.org/10.1016/j.buildenv.2017.07.007

Dwaikat, L. N., & Ali, K. N. (2016). Green buildings cost premium: A review of empirical evidence. *Energy and Buildings, 110*, 396–403. https://doi.org/10.1016/j.enbuild.2015.11.021

Eichholtz, P., Kok, N., & Quigley, J. M. (2010). Doing well by doing good? Green office buildings. *American Economic Review, 100*(5), 2492–2509. https://doi.org/10.1257/aer.100.5.2492

Energy Information Administrative. (2020). *2020 RECS survey data*. Housing Characteristics Tables. https://www.eia.gov/consumption/residential/data/2020/#sh

Gabay, H., Meir, I. A., Schwartz, M., & Werzberger, E. (2014). Cost-benefit analysis of green buildings: An Israeli office buildings case study. *Energy and Buildings, 76*, 558–564.

Gou, Z., & Lau, S. S.-Y. (2014). Contextualizing green building rating systems: Case study of Hong Kong. *Habitat International, 44*, 282–289. https://doi.org/10.1016/j.habitatint.2014.07.008

Haapio, A., & Viitaniemi, P. (2008). A critical review of building environmental assessment tools. *Environmental Impact Assessment Review, 28*(7), 469–482. https://doi.org/10.1016/j.eiar.2008.01.002

Hirvonen, J. (2021). *Finland: Heat pump market outlook*. Technology Collaboration Programme. https://heatpumpingtechnologies.org/magazine-1-2021/finland-heat-pump-market-outlook/

Hu, M., Pelsmakers, S., Vainio, T., & Ala-Kotila, P. (2022). Multifamily building energy retrofit comparison between the United States and Finland. *Energy and Buildings, 256*, 111685. https://doi.org/10.1016/j.enbuild.2021.111685

Hu, M., & Skibniewski, M. (2021). Green building construction cost surcharge: An overview. *Journal of Architectural Engineering, 27*(4), 04021034.

Hwang, B.-G., & Low, L. K. (2012). Construction project change management in Singapore: Status, importance and impact. *International Journal of Project Management, 30*(7), 817–826. https://doi.org/10.1016/j.ijproman.2011.11.001

Liu, Y., Guo, X., & Hu, F. (2014). Cost-benefit analysis on green building energy efficiency technology application: A case in China. *Energy and Buildings, 82*, 37–46. https://doi.org/10.1016/j.enbuild.2014.07.008

Meron, N., & Meir, I. A. (2017). Building green schools in Israel. Costs, economic benefits and teacher satisfaction. *Energy and Buildings, 154*, 12–18.

Nguyen, B. K., & Altan, H. (2011). Comparative review of five sustainable rating systems. *Procedia Engineering, 21*, 376–386.

Paulson, B. C. (1976). Designing to reduce construction costs. *Journal of the Construction Division, 102*(4), 587–592. https://doi.org/10.1061/JCCEAZ.0000639

Ramboll. *Sustainable Buildings Market Study 2019*. Retrieved April 2022. https://ramboll.com/-/media/files/rgr/documents/markets/buildings/s/sustainable-buildings-market-study_2019_web.pdf?la=en

Shrestha, P. P., & Pushpala, N. (2012). Green and non-green school buildings: An empirical comparison of construction cost and schedule. In *Construction Research Congress 2012: Construction Challenges in a Flat World* (pp. 1820–1829).

Shwartz, M. (2005). *Boyd Paulson Jr., civil engineering professor, dies of cancer*. https://news.stanford.edu/news/2005/december7/paulson-120705.html

Sisson, P. (2021, October 26). *As risks of climate change rise, investors seek greener buildings*. https://www.nytimes.com/2021/10/26/business/climate-change-sustainable-real-estate.html?auth=login-google

Sovacool, B. K., Gilbert, A., & Nugent, D. (2014). Risk, innovation, electricity infrastructure and construction cost overruns: Testing six hypotheses. *Energy, 74*, 906–917. https://doi.org/10.1016/j.energy.2014.07.070

Tatari, O., & Kucukvar, M. (2011). Cost premium prediction of certified green buildings: A neural network approach. *Building and Environment, 46*(5), 1081–1086. https://doi.org/10.1016/j.buildenv.2010.11.009

Torcellini, P., Pless, S., & Leach, M. (2015). A pathway for net-zero energy buildings: Creating a case for zero cost increase. *Building Research & Information, 43*(1), 25–33. https://doi.org/10.1080/09613218.2014.960783

US Census Bureau Construction Spending Survey. (n.d.). *US Census Bureau Construction Spending Survey*. Retrieved June 19, 2021, from https://www.census.gov/construction/c30/methodology.html

World Green Building Council. (n.d.). *World Green Building Trends 2018 SmartMarket Report*. World Green Building Council. Retrieved June 19, 2021, from https://worldgbc.org/wp-content/uploads/2022/03/World-Green-Building-Trends-2018-SMR-FINAL-10-11.pdf

Zahirah, N., Abidin, N. Z., & Nuruddin, A. R. (2013). Soft cost elements that affect developers' decision to build green. *International Journal of Civil and Environmental Engineering, 7*(10), 768–772.

Zhang, Y. F., & Fuh, J. Y. H. (1998). A neural network approach for early cost estimation of packaging products. *Computers & Industrial Engineering, 34*(2), 433–450. https://doi.org/10.1016/S0360-8352(97)00141-1

2   **Existing research**

ABSTRACT

This chapter delivers a literature review on the sustainable building construction cost (SBCC) to provide a better understanding of relevant research activities and determine knowledge gaps and trends. To understand the SBCC, an overall understanding of the building construction cost is necessary. The author first outlines the research activities from 1980 to 2021 on the general building construction cost as well as identifying and describing knowledge gaps. This is followed by a review of the components included in a building's construction cost estimation and the terminologies used. The focus then switches to the SBCC, with its components and current research status.

NOMENCLATURE

AACE	Association for the Advancement of Cost Engineering
ASTM	American Society for Testing and Materials
BEES	Building for Environmental and Economic Sustainability
BREEAM	Building Research Establishment Environmental Assessment Method
CASBEE	Comprehensive Assessment System for Built Environment Efficiency
CIB	International Council for Research and Innovation in Building and Construction
LEED	Leadership in Energy and Environmental Design
SBCC	Sustainable building construction cost

2.1 RESEARCH DEVELOPMENT OF THE GREEN BUILDING CONSTRUCTION COST

Before 1980, there were few scientific research publications that focused on building construction costs, regardless of building construction cost estimations already being a relatively established field in practice. The earliest record of building construction cost data used in practice can be found in the online database

DOI: 10.4324/9781003316855-3

"Volume and Cost of Building Construction, 1914 to 1924," published by *Monthly Labor Review* (Byer, 1925). This type of cost data is similar to today's U.S. Census Bureau's publicly accessible data on monthly construction spending. From 1925 to 1980, only six publications were found on the topic of construction cost. Among these six publications, three focused on general building, one on school building, and one on apartment building. The remaining publication was titled "Energy Cost of Building Construction," published in 1977, which was the first known published study on the embodied energy cost of building construction in the United States (Stein, 1977).

Since the early 1980s, there has been an increase in research on building construction costs. As illustrated in Figure 2.1, the research can be divided into four periods: phase 1 (1980–1990), phase 2 (1991–2000), phase 3 (2001–2010), and phase 4 (2011–2021). The upper line represents publications of the conventional building construction cost, while the lower line represents publications of the green building construction cost.

2.1.1 Phase 1: 1980–1990

From 1980 to 1990, there was a steady increase in publications of the building construction cost, with a total of ten publications; most were from trade or professional organization publishers. Five out of ten publications were published in *Architectural Record*, which is the primary trade association publication for architects in the United States. Two studies were published as conference proceedings by CIB. The research topics focused on overall housing price fluctuations or construction cost forecasting rather than on cost factor analysis;

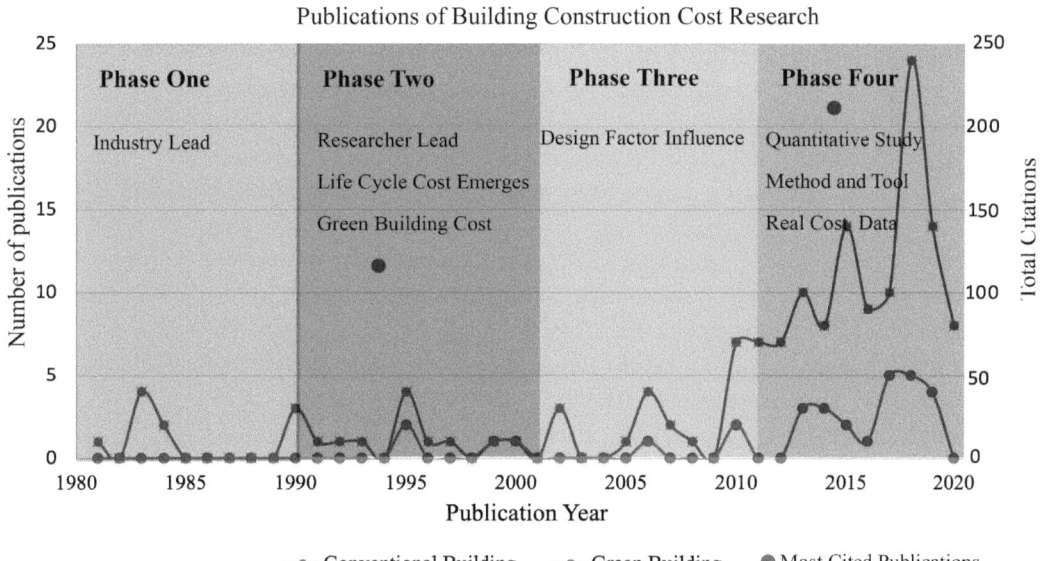

Figure 2.1
Publications of the building construction cost and green building construction cost

few studies were published as independent academic research. Interestingly, the two peak publication periods for construction cost studies aligned with the economic recession periods of 1981–1983 and 1990–1991. Since the building construction industry was deeply impacted by the recessions, there were needs and incentives for the industry and professional organizations to conduct their own studies and cost analysis. Overall, during this ten-year period, research activities around the building construction cost were dominated by the industry and trade organizations.

2.1.2 Phase 2: 1991–2000

From 1991 to 2000, 11 publications were found, which demonstrate two trends in research activities on the building construction cost. The *first trend* was the emergence of construction cost studies on green buildings. In the early 1990s, sustainable or green buildings were not well defined, so publications related to green building costs were often titled with the keywords "energy" or "CO_2." For example, one of the earliest studies, published in 1995, used a parametric cost model to study the construction cost increase and CO_2 emissions reduction by replacing non-sustainable building materials (cement) with sustainable building materials (lime) (Tiwari & Parikh, 1995). The researchers found that a 21% emissions reduction produced a 27% increase in building construction costs, which was mainly related to materials and labor costs (Tiwari & Parikh, 1995). The concept of life cycle cost savings from green building was also introduced in this period. In 1999, the *Journal of Construction Engineering and Management* published the paper "Selecting Cost-Effective Green Building Products: BEES approach," which is one of the most cited studies regarding the cost of green building products. In 2000, *Architectural Record* also published a short article introducing BEES software developed by the National Institute of Standards and Technology. BEES measures economic performance using the ASTM standard life cycle cost method, which covers the cost of the initial investment, replacement, operation, maintenance, and report and disposal of building materials (Lippiatt, 2000). BEES was developed in 1994, but it became known by the design community only in 2000, after the publication of the two articles.

The *second trend* was an increase in activity by academic researchers. Five of the eleven publications were by academic researchers, and six publications were by trade associations or practitioners. Most of the publications by academic researchers focused on the green building cost, while publications from the industry focused on cost estimating and advanced methods and tools. For example, parametric cost estimating of building construction was first introduced in a study presented at the Association for the Advancement of Cost Engineering (AACE) International's 37th Annual Meeting.

The main difference between research topics in phase 1 and phase 2 is the majority (eight out of eleven publications) of phase 2 research activities focused on construction cost factors and cost optimization/reduction.

2.1.3 Phase 3: 2001–2010

Phase 3 saw a steady increase in interest in building construction cost studies, but not in sustainable building construction costs (SBCCs). Out of a total of 18 publications, only three publications mentioned the SBCC. However, this period experienced the *most growth in green building rating systems*. Seven green building rating systems were established in seven countries across three continents. Logically, when a consumer product (e.g., LEED building) is introduced to the market, the performance and cost of the product is the most essential information consumers need to decide whether to buy. A green building is essentially a type of consumer product with a higher complexity and longer life span; therefore, there should been solid cost-benefit analysis before launching the product. However, such cost analyses **were not** done prior to launching those new green building rating system. Green building was apparently not treated as a consumer product by the early inventors and promoters; rather, it was promoted as a *lifestyle*, perceived as more environmentally responsive and potentially better than a conventional one. From the author's perspective, this **unaligned development** is the root cause of misperceptions of the actual construction cost of green building. Green building should not represent a type of building; it should signify a set of design and construction principles applied to all buildings. This concept will be further explained in detail in Chapter 7.

On the other hand, during this period, there was a tremendous increase in research from the academic community: of the 18 publications on building construction cost studies, 14 were by academic researchers. Consequently, most of the studies were based on hypothetical values or responses from surveys or questionnaires (Love, 2002) as academic researchers cannot easily acquire actual project construction cost data. This difficulty remains the primary obstacle for building construction cost-related studies. Due to the nature of academic research, many of the publications examined cost estimation methods and cost factors. Some of these design factors started to be included in studies. For example, in a study titled "Effect of building shape on a residential building's construction, energy and life cycle costs," the researcher found that the construction cost and energy use increased as the external wall area and floor area ratio increased. This finding suggests a strong correlation between design decisions and the construction cost.

Regarding geographic domination, interest and research activities during this period were mainly from Asia, Europe, and Australia, with little research in the United States that focused on the building construction cost. This trend continues to date, with **a lack of both interest and efforts** in US research on the building construction cost, which is reflected in the widespread ambiguity and misperceptions of the building construction cost, especially the SBCC.

One publication of interest during this period is "Time-cost relationships in Australian building construction projects." The researchers collected data through a survey of 161 building construction projects in Australia to study the relationship between time and cost. They found that new construction projects performed poorly compared to renovation projects, but they were not able to provide a logical explanation. They also concluded that the construction cost was a poor predictor of project time, and that the total floor area and number of floors were important determinants of construction time (Love et al., 2005). This study is critical since it disproved a long-standing time-cost model developed by Bromilow in 1969, which led to the development of a forecasting model for construction duration based on the cost (Bromilow, 1969). The non-linear relation between cost and time suggested that other types of correlation existed between time and cost; this indication is aligned with the effort curve proposed by Paulson (refer to Chapter 1).

2.1.4 Phase 4: 2011–2022

In 2011, there was a surge in interest in building construction cost research, especially in conventional buildings; however, interest in SBCC research remained relatively low. During phase 4, there were 111 publications on the conventional building construction cost but only 23 on the SBCC. From 2011 to 2022, a variety of cost factors were investigated besides the material and labor costs, with multiple regression being a commonly used technique (Lowe et al., 2006). Furthermore, there was some overlapping between conventional building construction cost research and SBCC research. One of the most cited papers, "Construction cost comparison between green and conventional office buildings," was published in 2013 (Rehm & Ade, 2013). The majority of such comparisons were based on expert interviews or surveys (Rehm & Ade, 2013), while real project construction data were still lacking.

Compared to the previous phases, two major research trends can be observed in phase 4. The first trend occurred following 2015: greater interest in improving research methods to gain a more quantitative understanding of how the construction costs of conventional and sustainable building were affected. Presumably, because of a lack of actual building construction cost data, there was a focus on a cost estimation comparison. There was also an exponential increase in exploration to integrate and introduce new tools, models, and methods in traditional construction cost estimations and analysis. Examples were a Monte Carlo simulation (Raydugin, 2017), stochastic annuity method (Fregonara & Ferrando, 2020), neural network model (Chandanshive & Kambekar, 2019), and BIM-based construction cost estimation (Akanbi et al., 2019), which were applied in construction cost estimations and analysis for sustainable and conventional building. The second trend was a shift within SBCC research: a shift in focus from the initial construction cost to building life cycle cost analysis. Among the SBCC publications published after 2015, 10 out of 13 focused on the life cycle cost.

In phase 4, the most promising trend in the past five years is the use of real project construction costs in studies. For example, Sun et al. (2019) obtained publicly accessible cost data for 37 green building-certified residential buildings and 36 general residential buildings in Taiwan. Their results showed the average construction cost of a sustainable residential building was only 1.58% higher than that of a general residential building, indicating that sustainable buildings are not necessarily more expensive than conventional buildings. However, they also found that achieving a high level of sustainability (certification level) is associated with an increase of 6.7% to 9.3% in construction costs. Further data from actual built projects would provide researchers and practitioners with in-depth knowledge of the most influential components and factors and would be imperative to eventually determine whether sustainable building is more costly.

2.1.5 Summary of research activities

Since the 1980s, there has been a substantial increase in research on the building construction cost, especially during 2010–2021, but growth in research on the SBCC has been limited. As illustrated in Figure 2.2, the primary research leader has switched from industry and practitioners to academic researchers.

The research topics are wide-ranging, from a new cost estimation method to avoiding cost overruns. **The primary research method has been conducting surveys and interviews, and the data have been mainly qualitative.** For decades, a lack of actual building construction cost data has been a barrier for research on the building construction cost. Moreover, the absence of data also contributes to a delayed interest in SBCC research since sustainability has been viewed as a lifestyle choice with a high price tag. **Consequently, the lack of quantitative studies is closely tied to the widespread perception of sustainable building being expensive, despite the lack of robust empirical evidence.** Another general knowledge gap observed is the design team's impact on the construction cost. The dominant focus has been on the construction method, labor/material supply, and construction management.

After gaining a comprehensive understanding of research on the building construction cost during 1980–2022, the author has concluded the main research gaps are an absence of quantitative data and a lack of studies on the design team's influence. In the following sections, the focus will shift to examine previous studies on the SBCC (in comparison to conventional buildings) to understand whether previous literature is sufficient for researchers to derive cost factors to

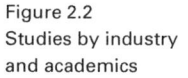

Figure 2.2
Studies by industry
and academics

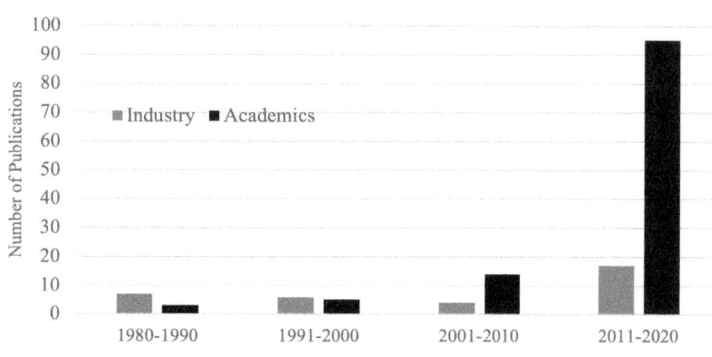

be included in further analysis. In addition, commonly used cost components will be explored, with the hope of gaining insight into why quantitative data are difficult to obtain.

2.2 CHANGES IN DEFINITION OF COST CATEGORIES AND COMPONENTS

A variety of terminologies are used for cost unit: first cost, capital cost, investment cost, direct cost, indirect cost, hard cost, and soft cost. The most used pairs of cost components are (1) direct costs and indirect costs and (2) hard costs and soft costs. Hard costs and soft costs are terminologies adopted for this study; in the following sections, the items included in these cost categories are examined.

2.2.1 Direct costs versus indirect costs

Carr was the first to clearly define direct costs as the costs that are traceable to physical activity (Carr, 1989); therefore, it is not a direct cost if an activity has not been performed (Houghton et al., 2009). Indirect costs are business costs other than the direct cost of construction activities; they are not physically traceable and are counted even if the activity is not performed. An indirect cost is also known as overhead (Houghton et al., 2009). AACE International defines direct costs as those resources that are expended solely to complete the activity or asset. The direct cost includes the cost of materials, labor, and equipment, while the indirect cost (overhead) consists of project overhead, general overhead, and other transactional costs. Project overhead includes the costs that are economically traceable to a project, such as a superintendent's salary and a tower crane rental. Small items and tools, such as nails and wires, are also considered project overhead. General overhead includes general office overhead, which is typically not traceable to construction activities or a project but is necessary for operating an office and obtaining work and other administrative work. It also includes general office and warehouse rentals, insurance, salary, and others (refer to Table 2.1).

2.2.2 Variable costs versus fixed costs

Another way to classify the construction cost is to differentiate between variable costs and fixed costs (Lowe et al., 2006). This is similar to a method used with accounting principles. Concrete work can be measured in square feet (meters), dry wall work in linear feet (meters), and site work in cubic yards (meters). Since these costs vary as the quantity of work changes, they can be considered variable costs. Fixed costs are costs that remain the same despite changes in work quantity and volume (Morris & Hough, 1987). For instance, the cost to rent a tower crane per day is fixed regardless of how much work is performed. Typically, project and general overhead – such as general office rentals, administrative costs, taxes, and salaries – are fixed costs. Both variable costs and fixed costs

Table 2.1 Breakdown of direct and indirect costs (adopted from the AACE manual, 6th edition)

Categories	Items	Subitems
Direct	Materials	
	Labor	Contractors
		Design professionals
	Equipment	
	Permit fee	
Categories	Items	Subitems
Indirect	Project overhead	Labor (e.g., superintendent's salary)
		Equipment (e.g., tower crane rental)
		Materials (e.g., nails, wires)
	General overhead	General office/admin. salary
		Office rental/warehouse rental
		Taxes/insurance
		Office supplies

can be direct or indirect costs. These terms were used to describe and analyze costs prior to more supplementary methods being introduced around 2005.

2.2.3 Hard costs and soft costs

Compared to other cost definitions, hard costs and soft costs were introduced to the building construction industry later and are more broadly defined. Geltner et al. (2001) defined hard costs as a type of direct cost of the physical components of a construction project, which include labor, materials, equipment, and overhead costs from the contractor and developer. Soft costs comprise the design, legal fees, and financing. Soft cost elements affect developers' decision to build green. They defined six soft cost elements: consultants, green building consultants, certification, commissioning, marketing, and taxes (Zahirah et al., 2013). Mapp et al. (2011) studied bank buildings and defined soft costs as the costs associated with the design, engineering, consulting fees, and certification fee, among others.

2.2.4 Initial cost and life cycle cost

The initial cost, or capital cost, has sometimes been used interchangeably with the construction cost. Fuller defined the initial cost as a capital investment that includes land acquisition, construction costs (or renovation), and the equipment costs needed to operate the building (Fuller & Crawford, 2011). Qian and Foong (2013) included all costs associated with procurement, supply, transport, and installation in the initial cost. In general, the initial cost includes the construction cost but goes beyond the costs associated with construction activities.

Life cycle cost analysis (LCCA) was first introduced to evaluate green building material and product studies. In 1995, the National Institute of Standards and Technology (NIST) published the *Life Cycle Costing Manual* (Handbook 135), which was developed for use in performing LCCA of investments in federal buildings and facilities (Qian, 2013). According to Handbook 135, the life cycle cost includes the initial investment, the annual cost for operation, maintenance, and repairs. In later research, building demolishing and deconstruction costs were included as well (Gopanagoni & Velpula, 2020). Despite the established life cycle cost method and manual, the adoption of and interest in life cycle cost estimation in the building and construction industry was negligible. Even to date, the practice of LCCA in construction projects is not common. Research interest in LCCA gradually increased in early 2010, mainly in green building construction costs. The lack of interest in life cycle costs from the construction industry is caused by split incentives. The developers, contractors, and design teams only bear the costs of and benefit from the initial capital investment. The life cycle cost benefit is more related to the building operators and occupants, who are typically not part of the team that makes the capital investment decisions.

2.3 KNOWLEDGE GAPS IN CONSTRUCTION COST AND ESTIMATION

There are two primary knowledge gaps in the construction cost of a building: (1) the misuse of cost and price, and (2) a lack of cost estimation transparency.

2.3.1 Misuse of cost and price

Most existing literature on construction costs does not clearly define the types of costs and often does not differentiate between the building costs and the resale price. Carlston (1952) draws a distinction between building prices and building costs by referring to the former as the market price for building work payable by a client and the latter as the costs incurred by a contractor in carrying out work. In general, the price always reflects some consideration of profits, while the term cost normally does not include profit. However, sometimes the price is used to derive the cost if the profit factor is known. For example, the U.S. Census Bureau estimates the construction cost of new single-family houses by using housing starts and/or sales data from its Survey of Construction (SOC). According to the SOC, the total cost of a new, private single-family home is obtained by multiplying the number of units by the average construction cost per unit. For units built by a developer (to be sold or rented), the average construction cost is the average sales price at the time of start multiplied by the factor 0.8424. This factor eliminates an estimate of the cost of "nonconstruction" items, such as raw land, marketing costs, closing costs, profit, and movable appliances (United States Census Bureau, n.d.). For a home built for an owner, the total cost is the average contract value at the start of construction multiplied by the factor 1.102 to eliminate "nonconstruction" items and add the value of land development not already accounted for. The national statistics for construction costs of residential units are derived from the sales price regardless of the type, which could lead to inaccurate information about the real construction cost. In addition, such complicated and varied calculation methods cause confusion, leading to difficulty in creating a fair comparison. For example, an article published in the *New York Times* titled "California's Plan to Make New Buildings Greener Will Also Raise Costs" (Penn, 2021) uses the sales price to backtrack the construction cost and assumes that the addition of solar panels (accounting for less than 5% of the total building construction cost) will be responsible for the sharp rise in sales prices of new single-family houses. The sales price increase is influenced by many factors, such as the housing demand and supply, market trends, and other unpredicted social and environmental conditions (such as the COVID-19 pandemic). It is not reasonable to attribute a 20% sales price increase to the additional solar panel costs, which account for less than a 5% increase. In addition, the author of the article ignored the long-term utility savings that homeowners can benefit from by using solar electricity, which has a relatively short payback period. The mismatch or misuse of the sales price to represent the construction cost is not only misleading but also greatly obstructs the promotion of building green.

2.3.2 Lack of cost estimation transparency

According to the Association for the Advancement of Cost Engineering (AACE) International, as listed in Table 2.2, there are five levels of estimation that are primarily associated with project maturity. The different levels of estimation entail different estimation methods. Generally, the estimation methods can be classified into two broad categories: conceptual and deterministic. As the level of project maturity increases, more project information becomes available, and the

Table 2.2 Construction cost estimation methods and levels (adopted from AACE International)

Estimated level	Maturity level of project (% of completed design)	Method	Expected accuracy	End usage (purpose of estimation)
1	65–100	Deterministic	−10% to +15%	Final bid/tender
2	30–75	Deterministic	−10% to +15%	Bid/tender
3	10–40	Deterministic and/or conceptual	−20% to +30%	Budget authorization
4	1–15	Conceptual	−30% to +50%	Feasibility study
5	0–2	Conceptual	−50% to +100%	Concept screening

estimating method tends to progress from conceptual methods (stochastic or factored model) to deterministic methods. There are two factors that differentiate conceptual methods from deterministic methods. The first major difference is that in conceptual methods, the data or factors used are not a direct measure of the building being estimated, while deterministic methods use data taken from the estimated building. Related to the data difference, conceptual models require significant effort to gather the historical data prior to cost estimation; then, the gathered historical data are used to develop factors and estimating algorithms. On the contrary, deterministic methods require a large effort during the estimation to measure the quantity and volume of the data taken from the actual building.

Conceptual methods are typically used for levels 4 and 5 and sometimes for level 3. The purpose of providing a conceptual estimation is to quickly determine an approximate potential project cost without a detailed design or clearly defined scope of work. It is also used by owners and developers to evaluate whether the project or investment can meet the financial threshold, enabling strategic decisions and establishing the project's preliminary funding. There is a wide range of methods and techniques used, such as the end-product unit method, project comparison method, physical dimensions method, and parametric method. The guidelines only provide high-level principles and instructions; there is no agreed upon method or techniques specified by AACE International. In addition, because the cost estimation database, tool, or method is normally the proprietary property of cost estimators or owners, there are rarely publicly available data or studies by practitioners that explain the cost and cost estimation. Overall, non-transparency and misuse of information contribut to the root causes of many misperceptions around the building construction cost.

2.4 CURRENT RESEARCH STATUS OF THE SUSTAINABLE BUILDING CONSTRUCTION COST

Regarding sustainable building development from the lenses of a sustainable building rating system, as mentioned in Chapter 1, there are various green building rating systems globally with different criteria and scoring systems. Table 2.3 lists the names of rating systems and their corresponding country of origin and year of establishment.

The five most recognized and adopted green building rating systems to define sustainability of the buildings are ranked as follows: (1) Building Research Establishment Environmental Assessment Method (BREEAM), (2) Leadership in Energy and Environmental Design – which is assumed to be the most used rating system currently (LEED) (Zhang et al., 2017), (3) Green Star, (4) Green Globes, and (5) Comprehensive Assessment System for Built Environment Efficiency (CASBEE). Seven core criteria are used across the different rating systems to measure sustainability. The criteria are energy, site, indoor environment, outdoor environment, material, water, and innovation. Beside the shared common evaluation criteria, the rating systems vary in priorities, scoring methods, weights,

Table 2.3 Green building rating systems

Label	Name	Country	Year
ASGB	Assessment Standard for Green Building	China	2006
BEAM BREEAM	Building Environmental Assessment Method Building Research Establishment Environmental Assessment Method	Hong Kong UK	1996 1990
CASBEE	Comprehensive Assessment System for Built Environment Efficiency	Japan	2001
CEPAS	Comprehensive Environmental Performance Assessment Scheme	Hong Kong	2002
CSH	Code for Sustainable Homes	UK	2006
EPRS	Estidama Pearl Rating System	Abu Dhabi	2010
GBI	Green Building Index	Malaysia	2009
GG	Green Globes	Canada	2000
GM	Green Mark	Singapore	2005
GS	Green Star	Australia	2003
GSAS	Global Sustainability Assessment System	Qatar	2009
IGBC	Indian Green Building Council	India	2013
LEED	Leadership in Energy and Environmental Design	US	1994

and evaluation requirements. For example, LEED, BREEAM, GG, and GS give the highest importance to the "Energy" category. However, among these four systems, "Energy" is the dominant category for LEED, while for BREEAM, GG, and GS, "Health and Wellbeing," "Water Efficiency," and "Indoor Environmental Quality" also have high importance (refer to Figure 2.3). For CASBEE, "Comfort and Safety" is the most important category, relating to Japan's geological characteristics in which seismic safety and security are critical. Most of the green building rating systems are voluntary. While this variety of systems has provided flexibility for designers and developers, it has also created ambiguity and a pick-and-choose mentality aided by such flexibility. The unintended consequences are that the public views sustainability as a choice, and achieving sustainability is just a matter of checking boxes. In addition, it is difficult to compare the construction cost under different green building rating systems. Therefore, the author has included all green certified buildings in the literature review to gain a comprehensive global perspective.

2.4.1 Literature selection
This book focuses on the SBCC; therefore, literature that did not specifically address the construction cost was excluded. The literature included was selected

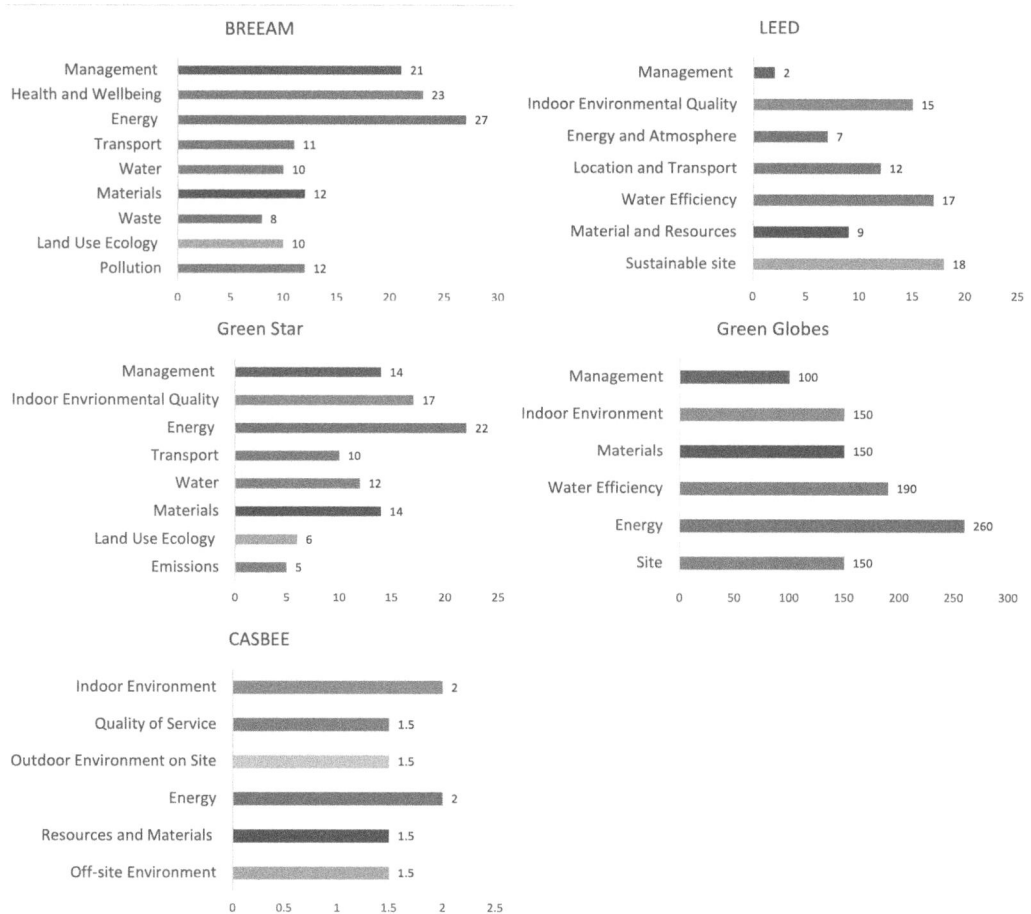

Figure 2.3
Categories and points for the five primary green building rating systems

based on the following criteria: (i) the publication addresses the SBCC as the main research topic, (ii) the publication relies on empirical data to draw a conclusion, (iii) the publication specifies the cost data resources, and (iv) the publication calculates the SBCC percentage, compared to that of the conventional building construction cost, where the percentage difference is defined as the sustainable building cost surcharge (SCS). The SCS is calculated using Equation 2.1:

$$\textbf{SCS} = \textbf{additional SBCCs / conventional building construction cost}$$
(Equation 2.1)

where *additional SBCCs* include hard and soft costs (refer to Section 2.3), and *conventional building* refers to those buildings not certified through any sustainable building rating systems. The buildings included in the previous studies differ in climate condition, function, construction type, location, and source of data. The data sources also vary widely, from questionnaires (surveys) to actual construction documents. Consequently, it is not appropriate to directly compare the cases against each other. The cases also differ in size and estimated lifetime. To neutralize these differences, the construction cost figures were normalized per unit of area ($/m²) and then compared in percentages of the additional SBCCs in relation to the conventional building construction cost.

2.4.2 Findings and conclusion

A total of 30 publications were included in this comprehensive review. As illustrated in Table 2.4, the SCS varies between −18.33% and 36.9%, and the empirical data used differs between studies published by academic researchers and industry experts. Some studies indicate that the cost of sustainable building involves high premium costs. Kim et al. (2014) revealed a 10.77% increase in the sustainable residential building construction cost, compared to that of conventional buildings not complying with green building codes. The researchers believed that the cost impact from the green building code was due to the incorporation of alternative power sources and energy-efficient appliances and equipment; the base building cost increase was not significant. In another study, Shrestha and Pushpala (2012) analyzed the construction costs and speed of construction of 30 green school buildings and compared the results to another 30 non-green schools. The authors comparatively concluded that the green school buildings cost 46% more than their conventional counterparts, and mean construction costs per square foot of the green schools were significantly higher than those of conventional schools. A case study was conducted by the National Association of Home Builders to itemize the cost impact of applying the National Green Building Standard to a green home. The findings determined that the cost premium was around 14% for a Silver-level certified house (Home Innovation Research Labs, 2014). The construction cost data used in the study were estimated data using the Residential Costs Book with RSMeans Data. The method they used to calculate the SBCC was to add the additional green costs to similar conventional buildings, which is guaranteed to result in a higher construction cost. The study did not consider that the added cost of increased insulation or energy-efficient fixtures could be offset by downsizing the mechanical equipment. Therefore, using the same conventional building cost as a base will not reflect the real cost of a sustainable building.

In contrast, other studies and findings indicate that green buildings can be achieved at a very low-cost premium. In the best-case scenario, there is no significant cost difference between sustainable buildings and conventional buildings. One of the most influential studies was commissioned by the Sustainable Building Task Force in California and conducted by a group of sustainable consultants as well as experts from a federal agency and national laboratory. They analyzed 33 LEED buildings, with analytic data showing that the median SCS was around 2%, or close to $4/ft², which is substantially less than what is generally perceived (Kats et al., 2003). The study also concluded that sustainable buildings cost less – on average, 30% less – to operate and maintain. In a later study, Kats (2006) investigated the design and construction costs of 30 green schools in different places in the United States and concluded that, on average, the SCS of green schools was around 2%, compared to conventional schools. Another important study that has been cited widely was conducted by a professional cost consulting company, Davis Langdon. They conducted studies in 19 different countries and collected data sets of almost 600 buildings. The building types included academic buildings, classrooms, laboratories, offices, hospitals, libraries, theaters, sports facilities, and museums. Their conclusion was that many projects achieved the sustainable goal without adding additional costs to their initial budget (Matthiessen & Morris, 2004). Another study carried out by Davis Langdon was a point-by-point analysis of 61 LEED-seeking projects. They found that those 61 projects were able to meet the LEED certification without additional costs. In a later study of a total of 221 buildings (with 83 of them being LEED-seeking projects and the other 138 buildings not having such sustainable goals), the researchers concluded that many projects were achieving LEED within their budgets and in the same cost range as non-LEED projects. The study also determined that the main obstacle to promoting sustainable building was not cost but **the idea that green is an added item** (Mathiessen & Morris, 2007).

The results from the literature review show a wide range in the cost variation of sustainable buildings, which does not provide an adequate quantitative indication of the cost components and whether there is a definite additional cost for sustainable buildings. This variation reflects the heterogeneity of the building stock. There are many factors that could cause these differences, such as building use (e.g., retail, office, hotel, education), building location (e.g., labor cost, material availability), climate and site conditions, and technologies and systems used in the buildings. Since the current literature review is not sufficient, a meta-analysis is needed to clearly understand the cost components and additional factors of sustainable buildings compared to conventional buildings. The analysis also needs to provide a breakdown of building types, locations, and cost data resources as well as information on who conducted the analysis. Based on this summary, a quantitative meta-analysis was conducted by the author, and the results are explained in Chapter 3's empirical findings.

2.4.3 Primary issue

Currently, there is no consensus on an SBCC definition (Houghton et al., 2009; Dwaikat & Ali, 2016). Also, there is no clear methodology describing sustainable buildings' additional construction costs compared to conventional buildings (Dwaikat & Ali, 2016). Data availability and transparency have been identified as primary barriers to studying construction costs, particularly sustainable construction costs (Chegut et al., 2019). A lack of clear methodology describing SBCCs is also identified as the leading cause for ambiguity and the large variation in costs of sustainable construction (Chegut et al., 2019). Some scholars have

Table 2.4 Sustainable building construction cost-related publications and data sources

Publication	Country	Year	Case study #	Type of building	Area (m2)	SCS (%)	Data source	Conducted by
Xenergy and Sera Architects	US	2000	1–3	OF	230–1,432	−0.3 to −1.3	Actual cost data from project team	Industry
Kats et al.	US	2003	33	SH, OF	NA	−0.66 to −6.5	Survey response (primary architects)	Industry
Matthiessen and Morris	US	2004	93	AC, LB	NA	No significant difference	Cost estimator database	Industry
Steven Winter Associates (GSA)	US	2004	2	OF	26,200–30,600	−0.4 to −8.2	Internet search	Industry
Bradshaw et al.	US	2005	16	RES	400–7,390	−18.33 to −8.15	Literature review and internet search	Industry
Kats	US	2006	30	SH	NA	0–6.27	Actual cost data from project team	Industry
Green Building Council Australia	Australia	2006	8	OF	NA	12–22	Australian Green Building Council private database	Industry
Matthiessen and Morris	US	2007	221	AC, LB, SH, OF	NA	No significant difference	Cost estimator database	Industry
Davis Langdon	Australia	2007	3	OF	>15,000	3–11	Cost estimator database	Industry
Construction Industry Institute	China (Hong Kong)	2008	38	OF	NA	0–3.2	Construction cost handbook (China & Hong Kong 2006) and cost estimator database	Industry
Fullbrook and Woods	New Zealand	2009	4	AC, HO, SH, OF	3,000	1.25–6.23	Actual cost data from project team	Industry
Houghton et al.	US	2009	13	HA	2,800–47,000	0–5	Survey response and interview	Academia
Kats et al.	US	2009	155	RES, HA, SH, LAB, others	240–200,000	0–18	Survey response (primary architects)	Industry
Mapp et al.	US	2011	10	Bank	281–512	<2	Actual cost data from project team	Academia

(Continued)

Table 2.4 (Continued)

Publication	Country	Year	Case study #	Type of building	Area (m2)	SCS (%)	Data source	Conducted by
Zhang et al.	China	2011	3	Hotel, OF, RES	33,610–23,710	8.5–13.9	Actual cost data from project team	Academia
Shrestha and Pushpala	US	2012	60	SH		No significant difference	Actual cost data from project team (Clark County School District)	Academia
Nyikos et al.	US	2012	160	All	NA	0.66–6.5	Online search	Academia
Rehm and Ade	New Zealand	2013	17	OF	NA	No significant difference	Actual cost data from project team	Academia
Building Research Establishment (BRE)	UK	2014	3	OF	13,800	0–5.7	Survey response	Industry
NAHB Research Center	US	2014	6	RES	232	0.6–14.65	Survey response	Industry
Kim et al.	US	2014	2	RES	270	10.77	WinEstimator database and 2011 RSMeans estimator	Academia
Gabay et al.	Israel	2014	6	OF	1,000–10,000	4.4–11.6	Israel Central Bureau of Statistics	Academia
Alexeew et al.	India	2015	4	RES	NA	0.8–2.8	Modeled cost	Academia
Garcia et al.	US	2015	8	RES	95–248.3	13.7–36.9	Modeled cost	Academia
Dwaikat and Ali	Global	2016	17	All	NA	−0.4 to −21	Literature review	Academia
Hwang et al.	Singapore	2017	363	All	NA	−4.5 to −7	Survey response (from project design)	Academia
Yih Chong et al.	Malaysia	2017	6	RES	NA	2.2	Modeled cost	Academia
Chegut et al.	UK	2019	542	All	NA	6.5	Royal Institution of Chartered Surveyors BCIS database	Academia
Sun et al.	Taiwan	2019	74	RES	2,085–122,097	−19 to −9.3	Government public information websites and architectural professional magazines	Academia
Ade and Rehm	New Zealand	2020	10	RES	NA	3–5	Author's own practice data	Academia

AC, Academic; OF, Office; SH, School; RES, Residential; HA, Health care; LAB, Laboratory.

tried to propose a framework to standardize the SBCC. For example, Houghton et al. (2009) defined the additional first-cost of green building as the additional construction costs associated with green design and construction elements but did not provide a detailed explanation of the meaning "first-cost." To provide a consistent and clear foundation for further investigation on the SBCC, the author proposed a construction cost components definition for sustainable building, which is explained in the following section.

2.5 PROPOSED COST COMPONENTS DEFINITION FOR SUSTAINABLE BUILDING

To provide a consistent definition and reliable comparison, in this book, the terms "hard costs" and "soft costs" are used. Although direct and indirect costs are the most used definitions in literature, the term "indirect" can be interpreted as unnecessary or unimportant. Indirect costs are often described as overhead, thus the project manager frequently tries to reduce and control overhead to make the project economically viable (profitable). For this reason, the author believes using the hard and soft cost definitions can avoid unnecessary bias.

In general, soft costs, such as the design cost, only account for a small portion of the total construction cost, around 3%–5% (Kats, 2006), but they play a significant role since most soft costs are capital costs and are borne by the developers upfront. Therefore, they have an immediate impact on the decisions of developers and builders to build green (Langdon, 2007). In this book (refer to Table 2.5), hard costs refer to the cost items required and directly related to physical construction activities of sustainable building, and soft costs refer to items not directly related to physical activities but that are necessary for completion of the activities.

Table 2.5 Proposed construction cost components and definitions used in this book

Hard costs	
Material	
Labor	Contractor
	Design professional on-site
Equipment	
Overhead cost	Developer
	Contractor
Soft costs	
Professional service fee	Architect/engineer
	Sustainable consultant
Administration fee	General office/inspection/management

2.6 CHAPTER SUMMARY

This chapter first summarizes research activities on the building construction cost from 1980 to 2022, which was divided into four phases. The research and knowledge gaps were identified. After understanding the overall research activities of the building construction cost, the focus shifted to the current research status of the SBCC. Findings indicate there is no consensus on whether sustainable building is more expensive than conventional building. In addition, the existing body of knowledge did not provide an adequate indication of the cost factors that are influential to the SBCC. Therefore, the author will conduct a meta-analysis of the SBCC and collect actual SBCC data to determine the root causes of the perceived higher construction cost of sustainable buildings. Finally, the terminologies of the cost components used in this book, soft costs and hard costs, were defined.

REFERENCES

Akanbi, T., Zhang, J., & Lee, Y.-C. (2019). Automated Item Matching and Pricing (IMP) for wood building elements to support BIM-based wood construction cost estimation. *Computing in Civil Engineering 2019*, 402–409. https://doi.org/10.1061/9780784482421.051

Bromilow, F. J. (1969). Contract time performance expectations and the reality. *Building Forum, 1*(3), 70–80.

Byer, H. B. (1925). Volume and cost of building construction, 1914 to 1924. *Monthly Labor Review, 21*(1), 173–179.

Carlston, K. S. (1952). Theory of the arbitration process. *Law and Contemporary Problems, 17*(4), 631. https://doi.org/10.2307/1190383

Carr, R. I. (1989). Cost-estimating principles. *Journal of Construction Engineering and Management, 115*(4), 545–551. https://doi.org/10.1061/(ASCE)0733-9364(1989)115:4(545)

Chandanshive, V., & Kambekar, A. (2019). Estimation of building construction cost using artificial neural networks. *Journal of Soft Computing in Civil Engineering, 3*(1), 91–107. https://doi.org/10.22115/scce.2019.173862.1098

Chegut, A., Eichholtz, P., & Kok, N. (2019). The price of innovation: An analysis of the marginal cost of green buildings. *Journal of Environmental Economics and Management, 98*, 102248.

Dwaikat, L. N., & Ali, K. N. (2016). Green buildings cost premium: A review of empirical evidence. *Energy and Buildings, 110*, 396–403. https://doi.org/10.1016/j.enbuild.2015.11.021

Fregonara, E., & Ferrando, D. G. (2020). The stochastic annuity method for supporting maintenance costs planning and durability in the construction sector: A simulation on a building component. *Sustainability, 12*(7), 2909. https://doi.org/10.3390/su12072909

Fuller, R. J., & Crawford, R. H. (2011). Impact of past and future residential housing development patterns on energy demand and related emissions. *Journal of Housing and the Built Environment, 26*(2), 165–183. https://doi.org/10.1007/s10901-011-9212-2

Geltner, D., Miller, N. G., Clayton, D. J., & Eichholtz, P. (2001). *Commercial real estate analysis and investments* (Vol. 1, p. 642). Cincinnati, OH: South-western.

Gopanagoni, V., & Velpula, S. L. (2020). An analytical approach on life cycle cost analysis of a green building. *Materials Today: Proceedings, 33*, 387–390. https://doi.org/10.1016/j.matpr.2020.04.226

Home Innovation Research Labs. (2014). *Cost and stringency comparison of 2012 National Green Building Standard™ ICC 700–2012, LEED-H 2008, and LEED v4 for homes design and construction*. National Association of Home Builders, p. 50.

Houghton, A., Vittori, G., & Guenther, R. (2009). Demystifying first-cost Green building premiums in healthcare. *HERD: Health Environments Research & Design Journal, 2*(4), 10–45. https://doi.org/10.1177/193758670900200402

Kats, G. (2006). *Greening America's Schools: Cost and benefit*. https://usd116.org/files/facilitiesreport/rptgreening.pdf

Kats, G., Alevantis, L., Berman, A., Mills, E., & Perlman, J. (2003). *The costs and financial benefits of green buildings: A report to California's sustainable building task force*. https://noharm-uscanada.org/sites/default/files/documents-files/34/Building_Green_Costs_Benefits.pdf

Kim, J.-L., Greene, M., & Kim, S. (2014). Cost comparative analysis of a new green building code for residential project development. *Journal of Construction Engineering and Management, 140*(5), 05014002. https://doi.org/10.1061/(ASCE)CO.1943-7862.0000833

Lippiatt, B. (2000). What's the Buzz? Use BEES to design greener, lower-cost buildings. *Architectural Record*. https://tsapps.nist.gov/publication/get_pdf.cfm?pub_id=860039

Langdon, D. (2007). "The cost & benefit of achieving green buildings". Retrieved April 2022. https://fdocuments.net/document/the-cost-benefit-of-achieving-green-buildings.html?page=3

Love, P. E. D. (2002). Influence of project type and procurement method on rework costs in building construction projects. *Journal of Construction Engineering and Management, 128*(1), 18–29. https://doi.org/10.1061/(ASCE)0733-9364(2002)128:1(18)

Love, P. E. D., Tse, R. Y. C., & Edwards, D. J. (2005). Time–cost relationships in Australian building construction projects. *Journal of Construction Engineering and Management, 131*(2), 187–194. https://doi.org/10.1061/(ASCE)0733-9364(2005)131:2(187)

Lowe, D., Emsley, M., & Harding, A. (2006). Predicting construction cost using multiple regression techniques. *Journal of Construction Engineering and Management, 132*(7), 8.

Mapp, C., Nobe, M., & Dunbar, B. (2011). The cost of LEED—An analysis of the construction costs of LEED and non-LEED banks. *Journal of Sustainable Real Estate, 3*(1), 254–273. https://doi.org/10.1080/10835547.2011.12091824

Matthiessen, L. F., & Morris, P. (2004). *A comprehensive Cost Database and Budgeting methodology*. Davis Langdon. https://vgbc.vn/wp-content/uploads/2018/12/Costing-Green_A-Comprehensive-Cost-Database-and-Budgeting-Methodology.pdf

Morris, P. W. G., & Hough, G. H. (1987). *The anatomy of major projects: A study of the reality of project management*. Wiley.

Penn, I. (2021, August 30). California's plan to make new buildings Greener will also raise costs. *The New York Times*. https://www.nytimes.com/2021/08/30/business/energy-environment/californias-solar-housing-costs.html

Qian, A. Y. (2013). A cost management approach to sustainable construction: Maximizing value via cost engineering techniques. *Proceedings of the SB, 235*, 42.

Qian, A., & Foong, W. (2013). A cost management approach to sustainable construction: Maximizing value via cost engineering technique. *Proceedings of the SB, 235*, 42.

Raydugin, Y. (Ed.). (2017). *Handbook of research on leveraging risk and uncertainties for effective project management*. IGI Global. https://doi.org/10.4018/978-1-5225-1790-0

Rehm, M., & Ade, R. (2013). Construction costs comparison between 'green' and conventional office buildings. *Building Research & Information, 41*(2), 198–208. https://doi.org/10.1080/09613218.2013.769145

Shrestha, P. P., & Pushpala, N. (2012). Green and non-green school buildings: An empirical comparison of construction cost and schedule. *Construction Research Congress 2012*, 1820–1829. https://doi.org/10.1061/9780784412329.183

Stein, R. G. (1977). Energy cost of building construction. *Energy and Buildings, 1*(1), 27–29. https://doi.org/10.1016/0378-7788(77)90007-X

Sun, C.-Y., Chen, Y.-G., Wang, R.-J., Lo, S.-C., Yau, J.-T., & Wu, Y.-W. (2019). Construction cost of green building certified residence: A case study in Taiwan. *Sustainability, 11*(8), 2195. https://doi.org/10.3390/su11082195

Tiwari, P., & Parikh, J. (1995). Cost of CO_2 reduction in building construction. *Energy, 20*(6), 531–547. https://doi.org/10.1016/0360-5442(94)00084-G

Zhang, Y., Wang, J., Hu, F., & Wang, Y. (2017). Comparison of evaluation standards for green building in China, Britain, United States. *Renewable and Sustainable Energy Reviews, 68*, 262–271. https://doi.org/10.1016/j.rser.2016.09.139

3   Meta-analysis findings

ABSTRACT

Chapter 3 discusses the meta-analysis findings on the sustainable building construction cost surcharge (SCS). First, the findings of the additional sustainable building construction cost (SBCC), compared to conventional buildings, on a global scale are discussed. Then, the cost factors related to the additional cost extracted from the meta-analysis are grouped into two categories: project characteristics and project team characteristics. Under each category, the variables and findings of how they influence the construction cost are summarized. The cost factors introduced in Chapter 3 serve as a foundation for the data analysis introduced in Chapter 4.

3.1 META-ANALYSIS RESULTS

As indicated from the preliminary literature review results summarized in Chapter 2, empirical research growth on the sustainable building construction cost (SBCC) is limited, mainly due to data availability. In addition, there has been *no meta-analysis* conducted on the sustainable building construction cost surcharge (SCS) on a global scale. Therefore, the author conducted a meta-analysis surveying the existing body of literature to aggregate the findings of empirical evidence that address the SBCC and to comparatively analyze the differences across building types, regions, and data sources to gain a deeper understanding of the SBCC. The SCS, as defined in Chapter 2, is used to measure the construction cost difference between conventional building and sustainable building (refer to Chapter 2, Equation 2.1).

Figure 3.1 shows a total of 31 studies, including more than **1,320** buildings from 11 countries, included in the meta-analysis. The projects span four continents: Asia, Europe, North America, and Australia. Primary cost factors are extracted from the meta-analysis findings to create a structural equation model (SEM), an explanatory framework explaining the SBCC, which can be found in Chapter 4.

3.1.1 *Differences in the SCS per building type*
Figure 3.2 illustrates that among the different building types, school (K-12) buildings have the highest mean (average) SCS, at 18%, which is much higher than all

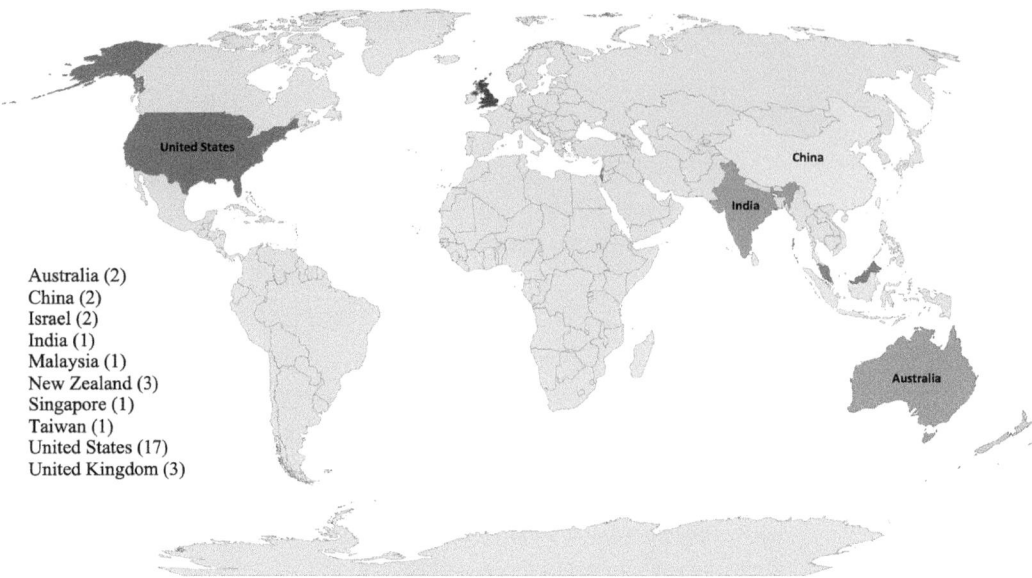

Australia (2)
China (2)
Israel (2)
India (1)
Malaysia (1)
New Zealand (3)
Singapore (1)
Taiwan (1)
United States (17)
United Kingdom (3)

Figure 3.1
Project location map

other building types. Office buildings have the second highest mean SCS, at 6%, while residential buildings have the third highest mean SCS, at 4%. Academic buildings (higher education buildings and other learning facilities) do not show a significant difference between the SBCC and the conventional building construction cost. In addition, most of the school and residential building costs are above the mean, whereas office buildings and other commercial building types are evenly split between cases with an SCS below and above the mean. The equal distribution of expensive and affordable sustainable commercial buildings represents a normal distribution of the construction cost, that maybe an indication that the knowledge and practices of sustainable office buildings have reached a certain maturity in which construction cost statistics are not skewed by a few extremely expensive or cheap projects. This maturity is well related to the dominant number of LEED-certified office buildings compared to other building types. These findings are aligned with previous studies where the *building type* was found to be a significant factor affecting the cost surcharge (AIA, 2020). However, the mechanism of how the building type impacts the SBCC has not been clearly explained.

Regardless of the difference, overall, based on empirical data collected from previous studies globally, the mean SCS is **6%** for all building types combined. In addition, the distribution between the projects above and below the mean are equal. Therefore, **6%** can potentially be used as a benchmark to describe the SCS across building types and regions.

Currently, in the United States, school buildings are leading efforts in advancing sustainable practices. According to a 2020 New Buildings Institute

Figure 3.2
Sustainable
building
construction cost
comparison per
building type

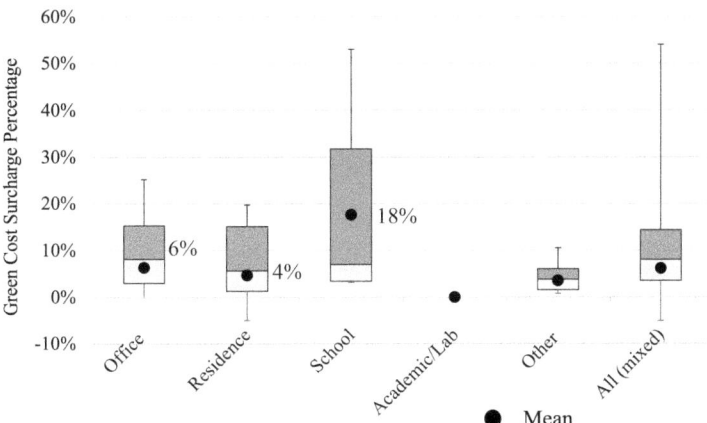

Figure 3.3
Zero energy
building list based
on NBI data: (a)
building type
breakdown, and (b)
education building
breakdown

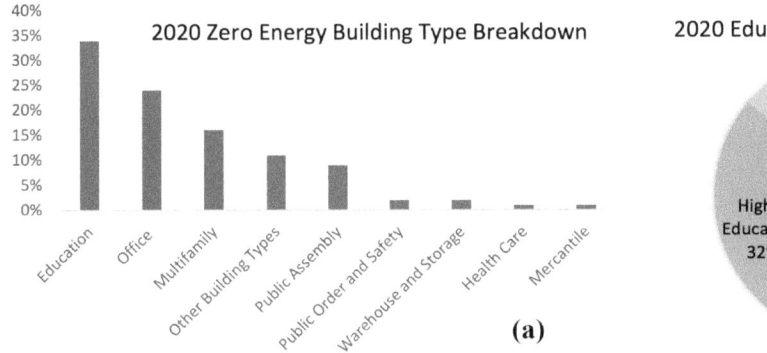

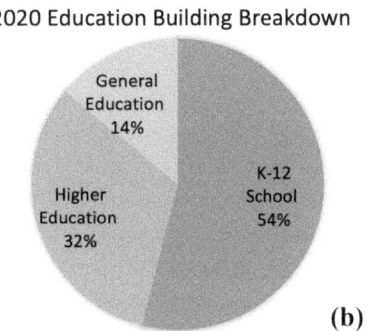

report, education buildings represented the largest portion of net-zero energy projects, at 34% (NBI, 2020) (refer to Figure 3.3a). Within education buildings, K-12 schools account for 54%, higher education buildings account for 32%, and general education buildings comprise the remaining 14% (refer to Figure 3.3b). School buildings, as a special type of public building, play a significant role in educating the public about and promoting sustainable building; thus the high SCS of net-zero school buildings can contribute to the public's perception of sustainable buildings being expensive.

3.1.2 Differences in the SCS per continent

The box plot in Figure 3.4a shows the mean SCS on a global scale: North America has the highest mean, at 7%, followed by Oceania (6%), Asia (5%), and Europe (3%). North America also has the largest SBCC variation among buildings, from −18.33% to 46%, while Europe has the smallest SBCC variation, from 0% to 6.5%, represented by the whiskers on top and bottom. The median value of the SCS in all regions is less than or equal to 6%, with Asia having the lowest value, at 2%, and Oceania having the highest value, at 6%. Although Asia

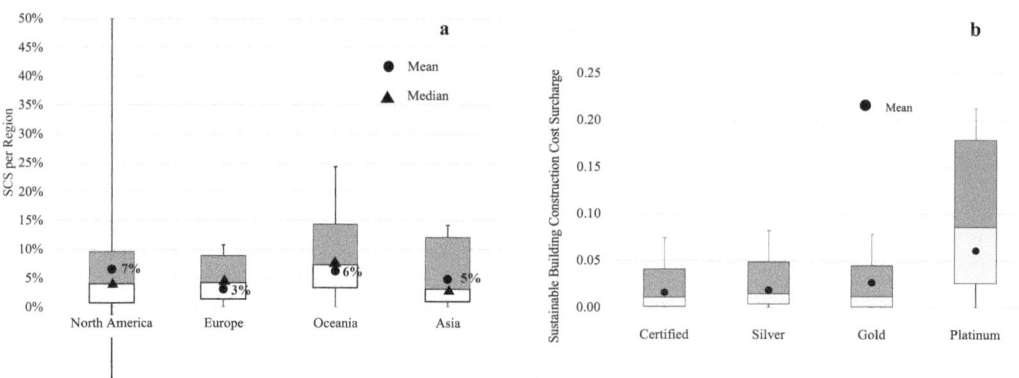

Figure 3.4
Sustainable building analysis (a) Sustainable building construction cost surcharge comparison across continents. (b) Sustainable building construction cost surcharge per LEED certification level

has the lowest median value, as illustrated in Figure 3.4, most projects are in the higher SCS range (>5%), which is represented by the gray boxes. Oceania also has more buildings in the high SCS range. North America and Europe have nearly an even number of projects in the higher and lower SCS ranges, which are represented by the equally sized gray and white boxes.

The wide range of cost variance in the North American projects is an indication of a lower level of maturity of sustainable building practices compared to Europe and a higher labor/material cost variation compared to Asia. Because of the higher average SCS and larger cost variance, further research on the North American market will provide in-depth knowledge on how sustainable building practices in North America can be improved to create a more affordable and less risky sustainable building market. Since most case projects included in North America are from the United States, Chapters 4 and 5 (Part II) will focus on this region.

3.1.3 Differences in the SCS per LEED certification level

The level of sustainability has been perceived as playing a role in the additional cost. The public assumes that a building with a higher level of sustainability equals a higher cost premium. According to existing empirical studies, Figure 3.4b demonstrates that the average (mean) SCS across different LEED certification levels is around **3%**. LEED Platinum buildings have the highest SCS, at 6%, while certified Silver and Gold buildings have an SCS of 2% and 3%, respectively. Additionally, LEED Platinum buildings have the largest cost variation.

Data show LEED-certified and Silver buildings' base construction costs were within the same overall cost range as that of conventional buildings. Further, added soft costs of LEED buildings were mainly due to administrative work and outside consultant fees, such as the LEED certification fee (Mapp et al., 2011). For LEED Gold and Platinum levels, different SCSs have been associated with individual projects; existing empirical studies did not provide further explanations or detailed information – only the cost premium in percentage terms.

For example, a General Services Administration-commissioned study showed LEED Gold buildings can increase the construction cost by 7% (Steven Winter Associates, Inc., 2004). A more recent study showed LEED Gold and Platinum certified buildings' additional construction costs were 7.43% and 9.43%, respectively (Uğur & Leblebici, 2018). In addition, most studies have not provided breakdown information on the level of sustainability of studied buildings; **hence it is not prudent to conclude from the meta-analysis that the level of sustainability is associated with the construction cost**. This is the problem the author sets out to explore in Chapter 4.

3.1.4 Differences in the SCS per construction cost data source

Among the included 31 studies, 22 studies (1,692 buildings) used actual building data, and the remaining 9 studies (77 buildings) used hypothetical data. Figure 3.5a illustrates the SCS statistics of actual buildings and hypothetical buildings. There are two types of hypothetical buildings: (1) the building does not exist and is created by researchers for study use, or (2) the building exists, but the real building information data cannot be obtained, so the researchers used estimated data and speculative information.

Two conclusions can be drawn. *First*, the average SCS for cases using actual building cost data is 6%, which is slightly higher than that of hypothetical buildings. However, the actual buildings' median SCS is lower than that of hypothetical buildings, by 2%. This indicates that the SCS of most actual buildings is higher than that of hypothetical buildings, due to the risk and uncertainties that occur during the construction. *Second*, actual buildings have a much larger cost variation, from −5% to 46%, which is associated with uncertainty and cost overrun during the actual construction process. Cost overrun is generally a symptom of inadequate planning and poor management (AIA, 2020). It was found that green building projects have higher cost overruns than conventional buildings (AIA, 2020). Using hypothetical buildings and modeled cost data may be sufficient to help the public gain an understanding of the average speculated green building construction cost. However, to account for uncertainty in a real project and ensure the actual construction cost will be within the budget, using actual project cost data is critical since the modeled cost cannot provide an accurate picture of the challenges and uncertainties that occur during construction.

Related to construction cost data used in previous empirical studies, as demonstrated in Table 3.1, there was no standard process for how the cost data were collected. Only 21% of the studies were able to obtain the actual construction cost data and documents from the project team, and 25% of the studies relied on survey or questionnaire responses from the project team members (architect, interior designers, engineers) and client or developers. Some studies used publicly accessible data rather than contacting project teams. For example, Chegut et al. (2019) used the Royal Institution of Chartered Surveyors' BCIS database, Sun et al. (2019) used the Taiwanese government's public information websites, and Gabay et al. (2014) used data from Israel's Central Bureau of Statistics.

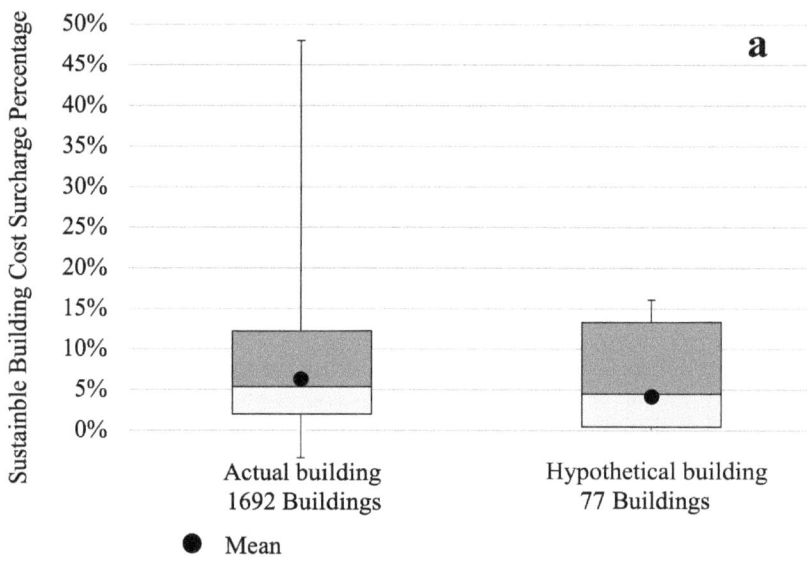

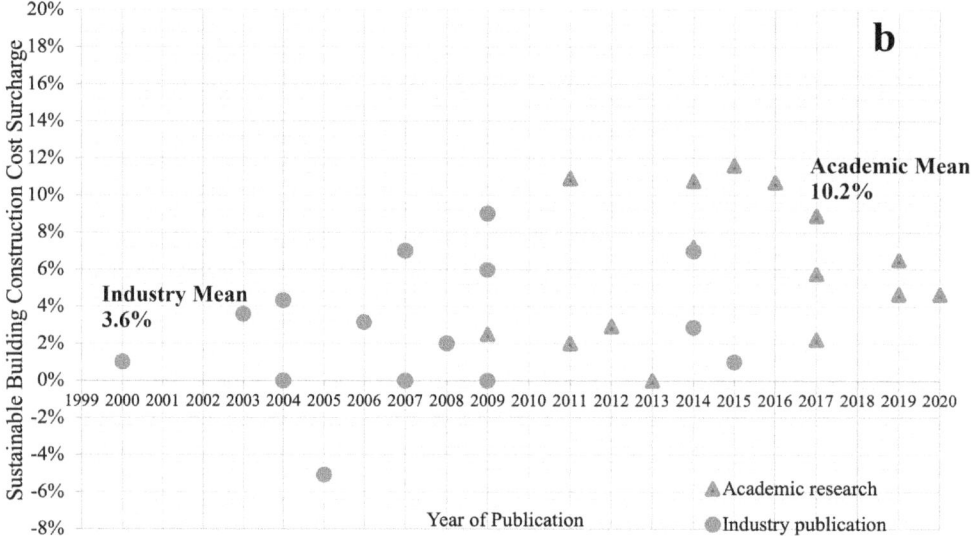

Figure 3.5

(a) Green building construction cost surcharge comparison between cost estimation methods. (b) Comparison of studies published by academic researchers and industry experts

3.1.5 Differences in the SCS between studies by industry experts and academic researchers

There are noticeable differences between studies published by academic researchers and those by industry experts. As illustrated in Figure 3.5b, 65% of the studies were conducted by industry experts and published by trade organizations, professional associations, or green building certification organizations. The data collection and research methods of these studies were typically not well defined or explained in the publications. Furthermore, the studies by industry

Table 3.1 Methods and cost estimation methods in prior studies

	Method	Number of Studies
Actual <p>building	Survey responses from professionals (cost estimators, contractors, architects)	5
	Actual cost information from project team (client, architects, engineers, cost estimators)	8
	Cost estimator database (Davis Langdon Database)	4
	Publicly available data (governmental databases or others)	4
	USGBC data (not accessible to the public)	1
	Author's own data from practice	1
Hypothetical building	Case studies of real projects using modeled costs and hypothetical green scenarios	4
	Model green building costs of actual building based on green specs	2
	Model green building costs of theoretical building based on design requirements	1
	Unknown	1

experts are older, with most of the studies published before 2010 and only three studies published after 2010. On the contrary, the majority of the academic publications are relatively recent, after 2010 (Dwaikat & Ali, 2016), and only one study was conducted and published before 2010. The mean SCS from the academic research is 10.2%, which is much higher than that of the industry-associated published studies, at 3.06%. It is understandable that in the early period, the primary cost data resources were gathered by industry experts, particularly professionals working in the green building field, and the research and reports based on empirical data demonstrating economic feasibility certainly have helped promote building green.

3.1.6 Summary of meta-analysis findings

Based on the identified construction cost differences between sustainable buildings and conventional buildings from the meta-analysis results, three summaries can be made.

First, it appears that the cost surcharge is smaller than the perceived value on a global scale, with a mean of 6%, and a median of less than 5%. However, school buildings have a much higher SCS, compared to other building types, which contributes to the perception of green building as expensive, especially since school buildings are leading the current efforts in building green.

Second, the data sources used for the studies influenced the analysis results: the studies using actual buildings had a higher mean SCS cost of 6%, compared to that of hypothetical buildings, at 4%. The varying procedures and data sources used in the studies clearly indicate that the standardization of data collection and validity are critical to reach consensus on the SBCC.

Third, there is a clear division of results, depending on who conducted the studies and for what purpose. Early studies conducted by practitioners showed small SCSs, while more recent studies from academic researchers had higher SCSs. The potential causes for such differences may be related to the study methods in addition to the data used.

Overall, the United States has a higher mean SCS and a large cost variance compared to other regions, thus it is beneficial to further investigate sustainable building in the United States. **The ambiguity and sometimes conflicting results of previous studies have fed into the perception of sustainable buildings' high construction cost.**

3.2 INFLUENTIAL COST FACTORS OF THE SBCC DERIVED FROM THE META-ANALYSIS

In this section, the cost factors for the SCS extracted from the meta-analysis are summarized and explained. The cost factors and variables will be used in Chapters 4 and 5 to generate a structural equation model to understand the causal relation between those factors and the total construction costs. Various factors have been studied in relation to the construction cost of both conventional and sustainable buildings. The author summarizes several important studies in chronological sequence below, in Table 3.2, to help readers gain a quick overview of the cost factors. These studies are among the most cited studies on cost factors that can be found in the last two decades.

Among the those studies, four are relevant to this book.

Lowe et al. (2006) studied 41 independent factors from data collected from 286 projects in the United Kingdom. They found those factors could be grouped into three categories: project strategic variables, design-related variables, and site-related variables. Among all variables, the gross internal floor area is the most significant cost-influential variable as it represents the *size* of the building. In addition, *building envelope*, *height*, and *project duration* are also well correlated with the construction cost.

Shane et al. (2009) studied 18 factors of construction projects and verified them by interviews with industry experts. These factors were categorized into an internal group (e.g., project schedule change and contract type) and an external group (e.g., inflation effects, market conditions, local environment, and unforeseen events). This study focused on understanding the factors through the literature review and focus group interview rather than identifying influential power and mechanisms, thus the researchers did not make any conclusions on the most influential factors.

Cha and Shin (2011) studied 49 cost factors in seven categories and identified their influence. The seven categories are schedule issues, budget issues,

Table 3.2 Most cited empirical studies on cost factors

Author, date	Cost category	Cost factors
(Akintoye, 2000)	1. Project complexity	1 Expected project organization 2 Type of structure 3 Site constraints 4 Method of construction and techniques 5 Scale and scope of construction 6 Complexity of design and construction
	2. Technological requirements	7 Amount of specialist work 8 Lead time 9 Off/on-site operations' sequence and limitations 10 Buildability
	3. Quality of information	11 Quality of information and information flow 12 Availability and supply of resources (labor & materials) 13 Precontract design 14 Expertise of consultants
	4. Project team requirements	15 Capacity of design team 16 Project team experience on the type of construction 17 Number of project team members
	5. Contract requirements	18 Type of client 19 Client's financial standing 20 Procurement route and contractual arrangement
	6. Project duration	21 Project duration 22 Anticipated frequency or extent of variation in construction requirements
	7. Market requirements	23 Project location 24 Tender period and market condition
(D. Lowe et al., 2006)	1. Project strategy	25 Contract type (DB, DBB)
	2. Design related	26 Envelope 27 Building type (function) 28 Building shape complexity 29 Gross floor area
	3. Site related	30 Site access 31 Brownfield vs. greenfield

(*Continued*)

Table 3.2 (Continued)

Author, date	Cost category	Cost factors
(Shane et al., 2009)	1. Internal	32 Bias 33 Delivery/procurement approach 34 Project schedule changes 35 Engineering and construction complexities 36 Scope changes 37 Scope creep 38 Poor estimations 39 Inconsistent application of contingencies 40 Faulty execution 41 Ambiguous contract provisions 42 Contract document conflicts
	2. External	43 Local concerns and requirements 44 Effects of inflation 45 Scope changes 46 Scope creep 47 Market conditions 48 Unforeseen events 49 Unforeseen conditions
(Cho et al., 2009)	1. Project characteristics	50 Time until budget is fixed 51 Percentage of repetitive elements 52 Project scope definition completion when bids are invited 53 Flexibility of work scope 54 Level of design preparation by owner 55 Time given to contractors to prepare bid
	2. Project participant characteristics	56 Owner's capability of construction management 57 Owner's administrative burden 58 Owner's experience with similar projects 59 Owner's level of control
	3. Environmental characteristics	60 Type of project 61 Level of construction complexity 62 Location of site
	4. Contractor characteristics	63 Project scale 64 Communication among project team members 65 Contractor's paid-up capital 66 Contractor's capability of construction

Author, date	Cost category	Cost factors
(Cha & Shin, 2011)	1. Schedule issues	67 Delay in procurement of materials 68 Local weather conditions
	2. Budget issues	69 National tax regulations
	3. Quality issues	70 Strong need to observe environmental laws and ordinance 71 Civil appeals caused by construction noise 72 Civil appeals caused by construction dust
	4. Safety issues	73 Not influential
	5. Environmental issues	74 Not influential
	6. Contractual issues	75 Vague contract terms cause disputes
	7. Managerial issues	76 Not influential
(Toh et al., 2012)	1. Project complexity	77 Type of construction 78 Construction method/technology 79 Project scale 80 Complexity of design and construction
	2. Project information	81 Level of adequate cost data 82 Available labor and materials 83 Project type 84 Floor area 85 Soil condition 86 Site topography 87 Complexity of building services 88 Quality of finishing
	3. Project team requirements	89 Previous records of claims and disputes 90 Contractor involvement in the design phase
	4. Market requirements	91 Material price fluctuation 92 Labor price fluctuation 93 Availability of equipment 94 Stability of market conditions 95 Tender period and market condition
	5. Technological requirements	96 Amount of specialist work 97 Quality of design and specs 98 Quality of construction required 99 Buildability of design

(Continued)

Table 3.2 (Continued)

Author, date	Cost category	Cost factors
	6. Contract requirements	100 Contract value 101 Type of contract 102 Financing and payment for complete work 103 Risk involved owing to the nature of the work 104 Client's financial situation and budget 105 Client requirements on quality
	7. Project duration	106 Duration of contract period 107 Quantity of expected variation in a project 108 Total factor productivity
(Rahman et al., 2013)		109 Fluctuation in price of materials 110 Contractor financial difficulties 111 Site management 112 Experience 113 Planning and scheduling 114 Contractor capability 115 Design drawing mistakes 116 Design changes
(Doloi, 2013)	1. Accurate project planning and monitoring	117 Planning and scheduling deficiencies 118 Methods/techniques of construction 119 Complexity of design and construction 120 Contractor's deficiencies in planning and scheduling at tender stage 121 Effective monitoring and feedback process
	2. Design efficiency	122 Extent of completion of precontract design 123 Mistakes and discrepancies in construction documentations 124 Client-initiated variations 125 Design changes within development period 126 Buildability
	3. Effective site management	127 Improper control over site resource allocations 128 Escalation of material prices 129 Cash flow during construction 130 Lower labor productivity 131 Delays in work approval waiting for information

Author, date	Cost category	Cost factors
	4. Communication	132 Lack of communication between client and contractor 133 Poor site management and supervision 134 Poor contract management 135 Lack of communication between design team and clients in design phase
	5. Contractor's efficiency	136 Inadequate contractor's experience 137 Low speed of decision-making 138 Project team's experience 139 Deficiencies in cost estimates 140 Contractor's financial difficulties
	6. Project characteristics	141 Scale and scope of project 142 Type of structure 143 Location of project 144 Unexpected geological conditions
	7. Due diligence	145 Understanding of responsibilities by all teams 146 Labor and management relations 147 Nonadherence to contract conditions
	8. Market competition	148 Tender period and market condition 149 Poor procurement programming of materials 150 Lead times for delivery of materials 151 Delay in subcontractor's work
(Cheng, 2014)	1. Environmental and circumstantial influence	152 Climate factors 153 Natural disasters 154 Geology, topography 155 High fluctuation in commodities
	2. Scope of contract	156 Clearly defined scope of project in contract 157 Modifications to scope of construction 158 Contract dispute (unclear drawings or guidelines/regulations) 159 Level of demand for quality
	3. Project risk	160 Gap between construction plan and reality is too great 161 Material shortage or supply delay

(*Continued*)

Table 3.2 (Continued)

Author, date	Cost category	Cost factors
	4. Management and techniques	162 Cost control 163 Project valuation does not match collected payment 164 Practical experience 165 Procurement contract 166 Time management 167 Project team (coordination capability and understanding of operational procedures)
(Obi et al., 2017)	1. Team quality	168 Competent design team professionals 169 Competent contractors 170 Team collaboration and commitment 171 Early contractor involvement
	2. Information and management	172 Detailed design documentation 173 Project planning and supervision
	3. Operational environment	174 Client finances 175 Stable economic environment 176 Weather conditions 177 Political environment
(Zhao et al., 2019)	1. Project characteristic factors	178 Project complexity 179 Supply chain 180 Site accessibility
	2. Key stakeholder's perspectives factors	181 Consultants 182 Client 183 Contractors 184 Building officers
	3. Market and industry conditions	185 Competition level 186 Resource markets 187 Market properties
	4. Regulatory factors	188 Building codes 189 Healthy and safety regulations 190 Construction contract act 191 Political policy 192 Financial regulations
	5. Macroeconomic dynamics	193 Global political stability 194 Natural disasters 195 Global economic trends 196 Labor cost

Author, date	Cost category	Cost factors
(Zhao et al., 2020)	1. Construction project system	197 Supply chain 198 Team experience and competence 199 Management relation 200 Organizational structure 201 Risk management 202 Contract type
	2. Economic-market climate	203 Market structure and size 204 Competition level 205 Economic stability 206 Investment management 207 Inflation 208 Interest rate 209 Credit supply condition 210 Exchange rate
	3. External environment	211 Political stability 212 Financial integration 213 Global economic climate 214 Natural disasters

quality issues, safety issues, environmental issues, contractual issues, and managerial issues. Among the variables from the categories, they found delay in procurement of materials, local weather conditions, national tax regulations, vague contract terms, civil appeal caused by construction dust and noise, and stringent environmental laws and ordinances were most likely correlated with project cost performance risk. Then, the research team used these influential factors to predict the project risk score for 12 real buildings. The test results revealed a statistically significant relationship between the project risk score and construction cost; hence these influential factors can potentially be used to predict the construction cost.

Doloi (2013) extracted 78 influential cost factors through a literature review; 48 of them were then used to develop a survey that was distributed to solicit responses from building industry stakeholders: clients, designers, consultants, and contractors. Next, the importance and influence of the 48 factors were grouped into eight factor groups; then, the importance of the groups and individual factors were tabulated and ranked. The data analysis showed the first three factor groups had a significant effect on the construction cost of buildings, which were accurate project planning and monitoring, design efficiency, and effective site management. The first group, accurate project planning and monitoring, can explain 23.9% of the cost variance among projects, which considerably emphasizes the technical competency of the project team. The second group, design efficiency, encompasses the significance of the extent of completion and quality of design prior to the onset of construction. It also counts the design changes during the construction. These first two factor groups can be described as project team characteristics. The third influential cost factor group, effective site management, explains 10.8% of the cost variance in projects, mostly referring to the influence of site-level management issues, such as material pricing, cash flow, and labor productivity.

Zhao et al. (2020) investigated 33 influential cost factors using a structural equation model based on a generalized maximum entropy and Bayesian estimation. Three categories were identified: construction project system, economic-market climate, and external environment. The construction project system (e.g., supply chain, team experience and competence, contract type, organizational structure) and economic-market climate (e.g., competition level, inflation, interest rate) have a significant effect on the construction cost, while the external environment does not.

3.2.1 Grouping of construction cost factors

As readers can see from Table 3.2 and the descriptions above, the factors, categories, definitions, and study methods used in the existing literature vary greatly, which has contributed to misperceptions and ambiguity of construction cost factors. Therefore, to help readers understand easily, this book categorizes factors affecting the construction cost that were identified in previous research into three groups: **project-specific characteristics** (e.g., building size and location), **project team characteristics** (e.g., the experience of design firm), and external characteristics (e.g., climate condition). Although the external characteristics play an important role in the construction cost, this book focuses on the first two groups.

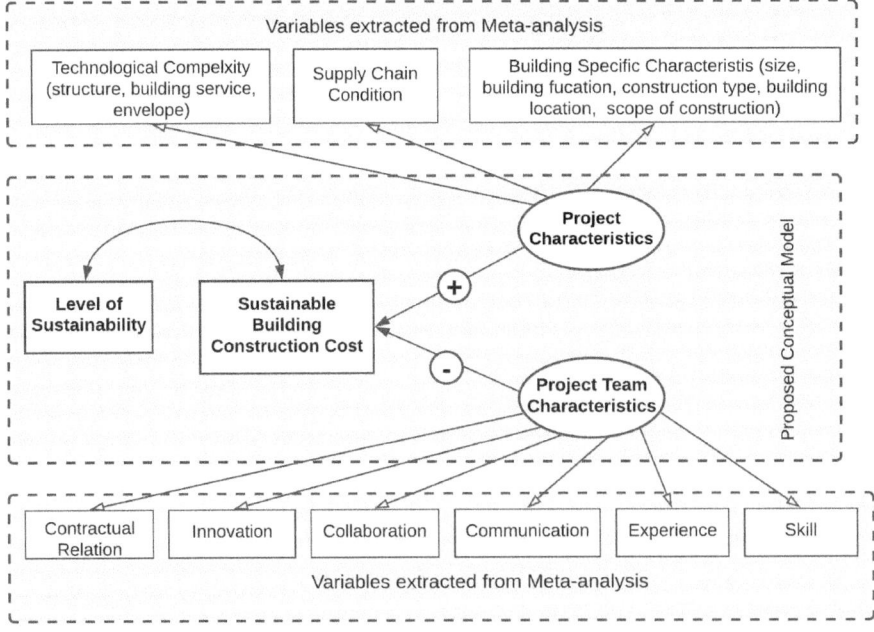

Figure 3.6
Cost influencing variables extracted from meta-analysis

Within these two groups, the criteria used to identify the characteristics from the meta-analysis are as follows: (1) data regarding the characteristics can be collected for the studied cases, and (2) the characteristics were identified in at least five previous studies. Certain characteristics that were mentioned in many studies but were difficult to collect for a large number of case projects were excluded from this study. For example, it is difficult to collect data on the labor productivity and project manager's working relationship with others for over 100 case projects that are included in data analysis in this book.

As illustrated in Figure 3.6, three subgroup project characteristics (technical complexity, supply chain condition, and building-specific characteristics) and six subgroup project team characteristics (contractual relation, team innovation, team collaboration, team communication, team experience, and team skill) were identified through the meta-analysis. In the following section, the author will explain how these characteristics influence the construction cost according to previous studies.

3.3 PROJECT CHARACTERISTIC VARIABLES

3.3.1 Technical complexity

The complexity in a construction project can be divided into two categories: organizational complexity and technological complexity. The organizational complexity is influenced by the organizational structure and relationships within a construction team (Raouf & Al-Ghamdi, 2019). The technical complexity for a construction project mainly concerns the building system, materials, construction method, and other issues related to transforming a design into reality. In this book, the author focuses on **technical complexity** since it is a commonly speculated factor and driver of the high SBCC. A complicated building system's direct impact on the construction cost is often reflected in higher material, labor, and equipment costs. It also indirectly impacts the construction cost through the demand for a more skilled design team, contractors, and a potentially longer design/coordination time and construction schedule.

In a study by Toh et al. (2012), among the 79 studied cost factors, complexity of design and construction was ranked as the second most influential factor to the cost (client requirement on quality was the most influential factor). Raouf and Al-Ghamdi (2019) stated the degree of technical complexity of sustainable buildings can be influenced by building control systems (e.g., mechanical and lighting systems), building energy systems (e.g., renewable energy systems), and building enclosure systems (e.g., exterior envelope systems) during the construction. Technical complexity can also arise in the design stage in sustainable building, mainly associated with the coordination among different disciplines with the aim to achieve a high building performance goal. This coordination happens among structural, architectural, electrical, security, communication, plumbing, and mechanical systems (Ahmad et al., 2016).

The definition of technical complexity is also related to spatial and temporal preference. For example, the heat pump (either air source or ground source) has been recognized as one of the most energy-efficient mechanical systems. It is widely promoted globally, with around 180 million heat pumps used for heating in 2020 and an increase in the global stock of nearly 10% per year over the past five years (2015–2020). However, the growth of the heat pump market has been uneven. The EU market is expanding quickly, with countries like Finland and Norway having the highest market penetration of around 25 heat pumps per 1,000 households each year (IEA, 2021). Conversely, in the United States, the heat pump is less common, with less than 40% of residential buildings using a heat pump system (EIA, 2018). The heat pump system, especially newer ground source heat pump systems, is not well known in the US building and construction industry. For buildings with high heating energy consumption, a ground source heat pump's energy-saving potential is larger than that of an air source heat pump (Häkämies et al., 2015). When a ground source heat pump system needs to be integrated with other building systems, US contractors are not equipped with the same level of experience as that of other countries. The difficulty of finding a skilled contractor leads to higher labor costs. This lack of knowledge and experience is directly linked to an increase in risk and uncertainty, consequently contributing to higher construction costs, a higher probability of cost overrun, and a higher probability of project schedule overrun.

Because of the wide range of technical complexity and location-specific issues, on a global scale, there is **no consensus on the most encountered technical complexity for sustainable building**. Therefore, the author must first determine the technical complexity in this book. As previously mentioned, the next logical step is to further investigate the SBCC in the United States. The most widely used sustainable building certification system in the United States is LEED. The LEED category Energy and Atmosphere has been found to be the most important and difficult to achieve full points. Therefore, the Energy and Atmosphere category in LEED was chosen as one of the measures for the technical complexity of sustainable building. In this LEED category, there are three prerequisite points and six possible credits related to energy performance, renewable energy, commissioning, refrigerant management, and measurement and verification.

3.3.2 Supply chain issues

In general, the supply chain condition may affect the availability of building materials and components and consequently the total construction cost. A mature supply chain can improve efficiency and productivity and reduce overall costs (Ofori, 2000). Conversely, supply chain delays have often been blamed as the cause for steep increases in material prices, alongside a labor shortage. Material-related issues contribute to project cost overrun (Koushki et al., 2005), and slow delivery of materials was identified as one of the top five factors causing project delays on a global scale (Sanni-Anibire et al., 2020).

The most recent evidence of the supply chain condition's profound impact on the construction cost was the supply disruption during the COVID-19 pandemic. The price of building materials and products for construction jumped nearly 20% in 2021, according to an analysis conducted by the Associated General Contractors of America (Supply Chain Dive, 2022). While the double-digit rise is notable in itself, several basic building materials spiked even more: lumber and plywood product prices increased by 62% and copper prices jumped 37% from April 2020 to February 2021. Due to the increasing material prices and labor costs, the construction cost of buildings rose by 17.5% between 2020 and 2021, which is the largest yearly increase since 1970 based on data from the U.S. Census Bureau's construction index (U.S. Census Bureau, 2022). Construction costs in 2021 increased by almost 23% compared to pre-pandemic 2019 levels (Quillen, 2022). According to Marc Hanson, the director of operations at CRB in Kansas City, "The amount of time it takes to get a shipment from China to the US has more than doubled from 35 days to around 78 days" (Yoders, 2021). In addition to the longer shipping time, the supply chain is also impacted by the port situation; a slow reopening due to the pandemic can create a large shipping backlog. For example, a record 62 cargo ships waiting to dock at the ports of Los Angeles and Long Beach, as of September 27, 2021, contributed to the supply chain crunch (Yoders, 2021).

During the pandemic, the SBCC was impacted by supply chain delays, labor shortages, and inflation, as were conventional buildings. However, the impact on the SBCC may be worse due to the immaturity and unsustainability of the supply chain for sustainable building materials. Even pre-pandemic, the higher cost of sustainable materials arose in part from the immaturity of the supply chain (AIA, 2020).

In developing countries, where the green building market is still new, certified green products or components must be imported, which also adds additional costs. Zhang et al. (2011) stated that the cost of sustainable materials is 3% to 4% higher than that of conventional building materials. Hwang and Tan (2012) showed how compressed wheatboard costs about ten times more than ordinary plywood. Amiril et al. (2017) also indicated how the shortage of sustainable building materials has been a barrier to promoting sustainable building in developing countries. In developed countries, such as the United States, the issues differ. For example, wood, as a commonly perceived "green" material, is loved by many designers, developers, and consumers in the United States. The recent steep increase in lumber prices is testimony to the demand of wood in the US market. According to the National Association of Home Builders (NAHB), the average 193 m² single-family house includes 13,127 board feet of framing lumber. In addition, plywood or oriented strand board, which covers the frame, exterior siding, trusses, doors, trims, flooring, sheathing, and moldings, are all made from wood products. Interestingly, the United States is a large producer of softwood lumber and certified wood (used for framing), milling over 26,200 billion board feet in 2003 and exporting over $380 million worth each year. However, the United States also imports about $4.5 billion worth of softwood lumber from Canada, over ten times as much as the United States exported to other countries. Furthermore, 13% of total imports comes from afar, including Chile, Brazil, New Zealand, Germany, and Sweden (Anderson, n.d.). The percentage of imported softwood lumber has increased in past decades; according to NAHB's 2021 report, the United States consumed roughly 47 billion board feet of softwood lumber in 2019, with 30.8% coming from 48 other countries (including Canada) (Logan, 2021). The high percentage of imported products explains why the supply chain had such a large impact on US lumber prices during the pandemic (2020–2022). **The concern is not whether the supply chain is mature enough but whether the wood products imported from thousands of miles away can still be considered sustainable materials.**

As mentioned in Chapter 1, when considering the embodied energy and carbon of building materials, imported materials have a much larger ecological footprint than conventional materials. From the author's perspective, the cost fluctuation caused by the supply chain in developed countries who heavily reply on imported products, like the United States, is not a maturity problem but a sustainability issue. Without a sustainable supply chain, a product cannot be considered sustainable, despite the material being perceived as and certified/labeled as a sustainable material. The most widely used and internationally recognized environmental label for sustainable materials is the Environmental Product Declaration (EPD), which is a report that provides information about a product's impact on the environment. EPD reports are generally available in the International EPD System's database, especially in Europe (EPD, n.d.). A report summarizing the environmental information of a product is normally fewer than 50 pages (EPD, n.d.). The International EPD System originated in the Nordic nations, where the first and current leading global EPD program was developed and operated in accordance with the ISO 14025, ISO 14027, ISO 14040 standards. For the building and construction sector, manufacturer EPDs also comply with the ISO 21930 and EN 15804 standards. Globally, there are 149 EPD products for the building and construction sector, which is a small number compared to the materials and products available for conventional building (Hu & Esram, 2021). The small amount of EPD building products may be related to the assessment cost. According to a cost survey conducted in 2022, the assessment cost for an EPD is between $13,000 and $41,000, and the workload is around 22 to 44 person-days (Tasaki et al., 2017). This high cost can hinder the adoption of EPD, especially for small and medium-sized manufacturers, who don't typically have the financial resources or manpower dedicated to obtaining an EPD label.

Similar to other green labels, EPDs also have their limitations. EPD and product category rules (PCRs) were developed to gather product data from the life cycle substages A1 to A3, which are associated with raw material extraction, transport, and manufacturing (The International EPD System, n.d.). However, the A4 and A5 substages are critical components of building construction and should not be neglected. Otherwise, low-carbon materials or products that are heavy, shipped over a long distance, or require energy-intensive equipment or processes to assemble on-site could be favored. For example, wood is a low-carbon construction material; however, its embodied carbon (in A4) may significantly increase as tree logs are shipped from their origin to a second location for primary processing (e.g., producing timber of a specific size and dimension), to a third location for secondary processing (e.g., making more specific building products), and to a distributor's warehouse before reaching the construction (Hu & Esram, 2021). The exclusion of the A4–A5 stages in the current EPD label indirectly creates the misperception of wood materials being sustainable regardless of their origin. The limitations of an EPD indirectly contribute to the perception of sustainable materials being expensive.

LEED is one of the sustainable building rating systems that has progressed the furthest in using EPDs to promote sustainable building materials. LEED v4 was introduced in 2013; in its current version (for new construction), there are two points available in the Materials and Resources category, which can be achieved by using at least 20 different permanently installed products with EPDs (one point), and through multi-attribute optimization (one point) (USGBC, 2021). Up to March 2022, there were only 511 certified projects under LEED v4, and not all projects had obtained an EPD point. Compared to a total of 17,416 LEED-certified new buildings (as of March 2022), 511 represents only 2.9%.

3.3.3 Building-specific characteristics

The building-specific characteristic category comprises the building type (function), project location, building geometry and complexity, scale and scope of construction, type of construction, site constraints, and other factors (Akintoye, 2000; Migliaccio et al., 2013).

The *project location* can have a direct impact on labor availability, equipment availability, local taxes, inflation, and other weather and climate conditions (Migliaccio et al., 2013). The location also has a considerable influence on the financial implication of construction projects. It is generally understood that costs in major metropolitan areas are higher than those in second-tier cities due to higher labor and material costs. Several researchers have specified location as a key cost driver on a global scale (Stoy et al., 2008; Dursun & Stoy, 2011).

Building geometry and *complexity* is defined by multiple attributes by different researchers (Morton & Jaggar, 1995). Kirkham et al. (2015) studied the building shape's impact on construction costs, with the included variables being the wall-to-floor ratio, plan compactness ratio, mass compactness ratio, length-breadth index, and plan shape index. Conversely, Ashworth and Perera (2015) defined building geometry as comprising the building height, plan shape, story height, grouping of buildings, and wall-to-floor ratio. Overall, there is no consensus on the most influential building-specific characteristics, and empirical studies on the relation between building-specific characteristics and the construction cost generally focus on cost estimation rather than on the actual cost. However, the building type and building size (floor area) are the two most studied factors (D. J. Lowe et al., 2006).

The *project size and scope* of work generally determine the overall cost and schedule of a construction project (Akintoye, 2000). Duy Nguyen et al. (2004) concluded that size (scope and scale of construction), as a single variable, is a determining factor to the success of construction projects. Bennett demonstrated that the size of a task affects the project cost breakdown and project duration. Regarding construction methods and types, scope of construction, construction technology, and types of construction were identified as the top three variables influencing cost estimations for conventional and sustainable buildings (Cho et al., 2009).

While sustainable building construction shares many similar cost factors with conventional buildings, such as location, size, and complexity, it also has distinctive factors. For example, prefabrication or modular construction, as a more sustainable construction method, can reduce construction waste (Tumminia et al., 2018). However, prefabrication housing was found to receive less demand due to its high cost and current cheap labor rate (Russ et al., 2018). In addition, a higher prefabricated building assembly cost was identified as one of the main reasons for a higher SCS (Mao et al., 2016). The construction methods or logistics also differ for sustainable and conventional buildings. Mao et al. (2016) indicated that in traditional construction, a tower crane enters the site once every nine months, whereas in some sustainable construction sites, a tower crane enters the site twice every eight months. Consequently, costs related to the tower crane contribute to the green cost surcharge (Home Innovation Research Labs, 2012).

3.4 PROJECT TEAM CHARACTERISTIC VARIABLES

3.4.1 Procurement and delivery method

The procurement and delivery methods have a direct influence on the selected project team characteristics and are often closely related. Several commonly used procurement and deliver methods are introduced below.

Procurement in the building and construction industry has been defined as a process of selecting and hiring a team for the design and construction of a project (El Wardani et al., 2006). The method of procurement is critical as it affects how a project will be delivered and financed (Hwang et al., 2017), impacts trust among stakeholders and openness and communication among team members (Shen et al., 2017), and influences risk management in construction projects (Osipova & Eriksson, 2011). There are four primary procurement methods for construction projects: low bid, best value (including fixed budget/best design), qualification-based, and sole source selection (El Wardani et al., 2006).

Sole source selection involves directly selecting a team without a proposal. Since the lack of competitiveness may discourage optimal results, it is not preferable unless the client has a time constraint that requires a shorter procurement method. *Qualification-based* selection is primarily based on the team's past performance, reputation, technical competence, and financial stability. In *best value* selection, the team submits a proposal, which is evaluated by the client based on technical aspects, associated costs, and other factors (Beard et al., 2001). With this selection method, clients may award the project to a team with whom they have established a long-term relationship and trust in technical capability. A *low bid* selection primarily selects the team based on the project value and related costs, where cost criteria represent more than 90% of the selection consideration (Beard et al., 2001). This selection method is characterized by a high level of design completion at the time of procurement to facilitate the competitive selection process (El Wardani et al., 2006).

The selection of the right team with effective organization is essential to ensure a successful project delivery without cost overrun or schedule overrun (Sanvido et al., 1992). In addition, the procurement method has been found to influence team culture building (Cheng, 2014). After deciding on the procurement method, the following step is to determine the project delivery method with the selected team. At times, decisions on the procurement method and delivery approach occur simultaneously.

The project delivery approach refers to the method and process used to deliver a construction project. The four commonly used project delivery approaches in North America are design-bid-build (DBB), design-build (DB), construction management (CM), and integrated project delivery (IPD) (Design-Build Institute of America, 2022).

DBB is a traditional method and has been the most widely used in both public and private construction projects. Further, project owners are the most familiar with DBB, and it is still required by some states (Kubba, 2012). In a DBB project, the owner first hires a design team to complete the design and then uses the design document (bid set) to secure a constructor who can commit to the project according to the design document within a certain budget and time. DBB usually provides the lowest construction cost, based on submitted tender documents, since the risk can be minimized through an owner's tight control. However, this method may take the longest to execute (Kubba, 2012).

DB is a method of project delivery in which the client hires one entity to be responsible for both design and construction. DB is usually the preferred contracting method under a tight schedule, and design-build contracts are often awarded through negotiation rather than through a bid process (Sweeney, 2022).

CM is also called CM at Risk, or simply CMAR, and is derived from the DBB process. Instead of the designer overseeing the design process and construction quality, a construction manager is hired by the owner to oversee the entire project. Under the CM delivery method, the construction manager acts as a client agent to ensure project delivery within a guaranteed maximum price and schedule (Zorich & Flipsen, 2018).

IPD is a relatively new delivery approach that integrates the project team to utilize the collective knowledge of all team members, reduce and manage risk, and optimize the project outcome. It is the highest form of collaboration because all participant parties are aligned by a single contract. This delivery approach is favorable for a sustainable building project that requires a high level of collaboration among clients, designers, and contractors from project conception to activation (Zorich & Flipsen, 2018).

Both decisions regarding which project delivery approach and procurement method to use affect the transfer of project risks. A lack of experience with a certain delivery approach or procurement method can lead to an underestimation of the project cost. Many agencies and owners look to reduce project risks and shorten schedules by using more integrated methods, such as DB or IPD. To find a qualified team who is familiar with DB or IPD and can execute the project, a qualification-based procurement method may be the most appropriate.

3.4.2 Skills and experience

Nearly all existing literature included the project team's skills and experience as influential factors to the construction cost. The expected project delivery capacity of the team is largely defined by the project team's experience with the type of construction and the number of similar projects the team has completed. Some studies focused on the client's experience and capacity (Chan et al., 2004) or the contractor's skills and experience (Cheng, 2014; Zhao et al., 2020), while others concentrated on the integrated team (designers and contractors) (Zhao et al., 2019) or the design team (Akintoye, 2000). With different foci, the definition of skills and experience varies.

Client-related factors include client characteristics, client type (e.g., public), client experience, and knowledge of the construction project's organization. The client's technical and decision-making skills are also influential factors to the project's success; however, technical skills are mostly related to financial estimation and project management rather than the technical knowledge of building systems and construction technologies (Chan et al., 2004).

The studies that focused on integrated team skills and experience did not concentrate on the client's skills, instead identifying the project manager as a vital player in achieving the project's success (Duy Nguyen et al., 2004; Nixon et al., 2012). Vrchota et al. (2020) conducted surveys at 114 companies, with the data analysis revealing the most significant factor for a project's success is an experienced project manager. Close to 50% of respondents ranked it as the top factor, whereas contradictory to common perception, financial resources was ranked much lower on the list. The project manager is not always the client; they often are a construction manager hired by the client, representing the client's interest. The manager/project leader's skills are wide ranging, including technical competencies (Heaton et al., 2016), communication or coordination skills (Sousa et al., 2018), planning skills, motivational skills, and organizing skills (Chan et al., 2004).

Studies focusing on the contractor's skills and experience also included similar skills, with practical experience and time management skills being important to contractors (Cheng, 2014). As the expected construction progress can be achieved only through the attainment of effective man-hours, a shortage of skilled personnel and related high labor costs and fluctuations in labor productivity were identified as major cost drivers in previous studies (Creedy & Kalb, 2005; Naoum, 2016).

The studies focusing on design teams found the designers' skills and experience regarding sustainable design to be important as they are related to technological complexity and uncertainty. Existing literature demonstrates the significant role of the project design team, spanning a project's inception to completion. Further, the accuracy and completion of construction documents produced by the design team directly relates to the frequency of changes that occur in the construction phase. Frequent design changes have been proven to lead to higher construction costs and longer construction schedules (Akintoye, 2000; Shane et al., 2009).

Two overall observations can be made from the existing literature. *First*, many studies focused on the skills and experience of the contractor or integrated team, while fewer studies investigated those of the design team and their related influence. *Second*, most data were collected using surveys or questionnaires. The responses are mainly based on the respondents' own experience and perceptions, rather than on empirical evidence. Consequently, to a certain extent, the identified factors and their related importance are the collective perceptions of the survey respondents. One recent trend identified from the studies indicates that the focus has shifted from individual players to the team. Regarding key success factors, studies found an effective team to be more important than individual project leaders (Obi et al., 2017).

3.4.3 Collaboration and communication

Communication is the process of acquiring all relevant information, interpreting this information, and effectively disseminating it to persons who may need it (Zulch, 2014). Communication is important for a project where more than one practitioner participates. Poor communication or a lack of communication between different stakeholders in a construction project is a common cause of increased construction costs and delayed schedules (Doloi et al., 2012). Communication in the building and construction industry can happen on two different scales: large and small. On a large scale, the information flows between different stakeholder teams, such as the client, contractor, design team, and consultants. On a small scale, the communication occurs between individuals within the same team (Gamil & Abdul Rahman, 2018).

In the construction industry, communication is managed by channels between all involved stakeholder groups (companies and firms). Due to complexity in the construction industry, severe communication problems occur because no proper channeling is adopted to manage and control the communication process (TIPILI & Ojeba, 2014). In addition to the proper communication channels and systems, communication tools and techniques are also identified as critical by the Project Management Institute. Commonly used examples include email, video-conferencing, and data sharing platforms (e.g., building information model) (Yap et al., 2018). The key to a project's success is effective and efficient communication with the correct people, at the right time, and in the appropriate format (Yap et al., 2018).

Collaboration is often referred to as working together; it can be reflected in different forms (Hughes et al., 2012) and is related to the procurement type. The definition adopted in this study was derived from a comprehensive review conducted in 2012, which is closely related to partnering (Hughes et al., 2012). Collaboration of a building project refers to the team effort by all participants at the early design stages, including the project manager, architects, engineers, consultants, owners, and potentially the construction managers. In previous studies on the green building cost, intensive upfront collaboration during the design stages was necessary to support technology application. For example, the project team of a multi-family apartment building (New York City) project claimed that the use of a single downsized boiler for heat and hot water was considered an innovative energy-saving technology and thus needed intensive design collaboration for implementation (Bradshaw et al., 2005).

3.4.4 Innovation

The construction industry is often compared unfavorably with the manufacturing sector's ability to generate technical innovation (Hardie & Newell, 2011). However, innovation is critical to long-term success in the construction industry. Within the construction industry, the definition of innovation provided by Slaughter (1998) is broadly accepted by academics and practitioners: "Innovation is the actual use of a nontrivial change and improvement in a process, product, or system that is novel to the institution developing the change." There is a wide range of innovation drivers and strategies in the construction sector from all stakeholders, including clients, designers, suppliers, contractors, end users, vendors and distributors, and certification bodies. Existing literature on construction sector innovation reveals that many forces drive firms to innovate to gain a competitive advantage. In general, competitive advantage refers to the ability of a firm (team) to perform at a higher level than others in the same market (Magretta, 2012). The drivers of innovation can be either technology-push or market-pull (Barrett & Sexton, 2006). More specifically, four push and pull drivers identified and described by Goffin and Mitchell (2017) are applicable to the construction industry: *technological advances*, *changing customers*, *intensified market competition*, and *changing business environments*.

The first driver, technological advances, normally originates within the industry. For example, in recent years, technological advancements, like 3D printing and robotic construction, have generated interest within the construction industry but have not been widely accepted or adopted in the market. The second, third, and fourth drivers are all market-pull drivers. The new requirements of clients—a need to develop new building performance standards in compliance with new energy efficiency regulations—have created momentum in the sustainable building market. As indicated in Chapter 1, investors eventually became aware of the financial benefits of building green. Client demand has become the top reason to construct sustainable buildings. According to the *World Green Building Trends 2021* report published by Dodge Construction Network, the top driver for building green is client or market demand, with clients wanting sustainable buildings due to the lower operating costs and improved occupants' health and well-being (Dodge Construction Network, 2021). The growing demand from clients and increasing market competition has stimulated innovation in the design and construction of sustainable buildings.

Conversely, according to a literature review and quantitative questionnaire surveys conducted by construction practitioners in 2017, technological advances ranked last among all drivers (Meng & Brown, 2018), illustrating that market-pull has a greater effect on driving innovation in construction than technology-push. These characteristics explain why construction is perceived as being low-tech compared to other industry sectors (Harty, 2008) and why construction innovation for sustainable building is **less driven by technological factors** and more by market factors and mindset shifts (perception). Put simply, the **market is a more dominant driving force for sustainable building construction.**

> Consequently, most innovation strategies are non-technical-driven, focusing instead on optimizing current operations and improving workers' skills. The *World Green Building Trends 2021* report also indicated that technological complexity and advancement are often **not** the primary obstacle to increasing sustainable building activities (Dodge Construction Network, 2021).

3.5 CHAPTER SUMMARY

Chapter 3 explains the findings from the meta-analysis. Although the meta-analysis results show a construction cost difference between sustainable building and conventional building, the difference of 6% (mean) is smaller than what has been perceived. Moreover, after more examination, the nontransparent and unstandardized data collection and sources method made the findings less robust and cast more doubt on their reliability. In addition, the clear division of findings between academic research and industry reports demonstrates the need for further examination. The second important aspect of Chapter 3 is that critical cost factors were extracted from the meta-analysis and organized into two categories: project characteristics and project team characteristics, which are used as a contextual foundation to generate a structural equation model (SEM).

REFERENCES

Ahmad, T., Thaheem, M. J., & Anwar, A. (2016). Developing a green-building design approach by selective use of systems and techniques. *Architectural Engineering and Design Management, 12*(1), 29–50. https://doi.org/10.1080/17452007.2015.1095709

AIA. (2020). *AIA issue brief: Green building rating systems legislation.* https://www.aia.org/resources/8801-aia-issue-brief-green-building-rating-systems

Akintoye, A. (2000). Analysis of factors influencing project cost estimating practice. *Construction Management and Economics, 18*(1), 77–89. https://doi.org/10.1080/014461900370979

Amiril, A., Nawawi, A., & Takim, R. (2017). The barriers to sustainable railway infrastructure projects in Malaysia. *The Social Sciences, 12*(5), 769–775.

Anderson, N. (n.d.). Where did your house grow? *ESF Suny College of Environmental Science and Forestry.* Retrieved June 27, 2022, from https://www.esf.edu/eis/eis-house-grow.php

Ashworth, A., & Perera, S. (2015). *Cost studies of buildings.* Routledge. https://doi.org/10.4324/9781315708867

Barrett, P., & Sexton, M. (2006). Innovation in small, project-based construction firms. *British Journal of Management, 17*(4), 331–346. https://doi.org/10.1111/j.1467-8551.2005.00461.x

Beard, J. L., Wundram, E. W., & Loulakis, M. C. (2001). *Design-build: Planning through development.* McGraw-Hill. https://www.accessengineeringlibrary.com/content/book/9780070063112

Bradshaw, W., Connelly, E. F., Cook, M. F., Goldstein, J., & Pauly, J. (2005). *The costs and benefits of green affordable housing.* https://www.newecology.org/wp-content/uploads/2017/08/The-Costs-Benefits-of-Green-Affordable-Housing.pdf

Cha, H. S., & Shin, K. Y. (2011). Predicting project cost performance level by assessing risk factors of building construction in South Korea. *Journal of Asian Architecture and Building Engineering, 10*(2), 437–444. https://doi.org/10.3130/jaabe.10.437

Chan, A. P. C., Scott, D., & Chan, A. P. L. (2004). Factors affecting the success of a construction project. *Journal of Construction Engineering and Management, 130*(1), 153–155. https://doi.org/10.1061/(ASCE)0733-9364(2004)130:1(153)

Chegut, A., Eichholtz, P., & Kok, N. (2019). The price of innovation: An analysis of the marginal cost of green buildings. *Journal of Environmental Economics and Management, 98*, 102248.

Cheng, Y.-M. (2014). An exploration into cost-influencing factors on construction projects. *International Journal of Project Management, 32*(5), 850–860. https://doi.org/10.1016/j.ijproman.2013.10.003

Cho, K., Hong, T., & Hyun, C. (2009). Effect of project characteristics on project performance in construction projects based on structural equation model. *Expert Systems with Applications, 36*(7), 10461–10470. https://doi.org/10.1016/j.eswa.2009.01.032

Creedy, J., & Kalb, G. (2005). Discrete hours labour supply modelling: Specification, estimation and simulation. *Journal of Economic Surveys, 19*(5), 697–734. https://doi.org/10.1111/j.0950-0804.2005.00265.x

Design-Build Institute of America. (2022). *What is design build?* https://www.dbiarockymountain.org/what_is_design_build.php#:~:text=Design%2Dbuild%20is%20a%20method,provide%20design%20and%20construction%20services

Dodge Construction Network. (2021). *World green building trends 2021.* https://www.corporate.carrier.com/Images/Corporate-World-Green-Building-Trends-2021-1121_tcm558-149468.pdf

Doloi, H. (2013). Cost overruns and failure in project management: Understanding the roles of key stakeholders in construction projects. *Journal of Construction*

Engineering and Management, 139(3), 267–279. https://doi.org/10.1061/(ASCE)CO.1943-7862.0000621

Doloi, H., Sawhney, A., & Iyer, K. C. (2012). Structural equation model for investigating factors affecting delay in Indian construction projects. *Construction Management and Economics, 30*(10), 869–884. https://doi.org/10.1080/01446193.2012.717705

Dursun, O., & Stoy, C. (2011). Time–cost relationship of building projects: Statistical adequacy of categorization with respect to project location. *Construction Management and Economics, 29*(1), 97–106. https://doi.org/10.1080/01446193.2010.528437

Duy Nguyen, L., Ogunlana, S. O., & Thi Xuan Lan, D. (2004). A study on project success factors in large construction projects in Vietnam. *Engineering, Construction and Architectural Management, 11*(6), 404–413. https://doi.org/10.1108/09699980410570166

Dwaikat, L. N., & Ali, K. N. (2016). Green buildings cost premium: A review of empirical evidence. *Energy and Buildings, 110*, 396–403. https://doi.org/10.1016/j.enbuild.2015.11.021

EIA. (2018). *EIA's residential energy survey now includes estimates for more than 20 new end uses*. https://www.eia.gov/todayinenergy/detail.php?id=36412

El Wardani, M. A., Messner, J. I., & Horman, M. J. (2006). Comparing procurement methods for design-build projects. *Journal of Construction Engineering and Management, 132*(3), 230–238. https://doi.org/10.1061/(ASCE)0733-9364(2006)132:3(230)

EPD. (n.d.). *The international EPD system*. Retrieved April 5, 2022, from https://www.environdec.com/about-us/global-house-of-https://www.environdec.com/home

Gabay, H., Meir, I. A., Schwartz, M., & Werzberger, E. (2014). Cost-benefit analysis of green buildings: An Israeli office buildings case study. *Energy and Buildings, 76*, 558–564. https://doi.org/10.1016/j.enbuild.2014.02.027

Gamil, Y., & Abdul Rahman, I. (2018). Identification of causes and effects of poor communication in construction industry: A theoretical review. *Emerging Science Journal, 1*(4). https://doi.org/10.28991/ijse-01121

Goffin, K., & Mitchell, R. (2017). *Innovation management: Effective strategy and implementation* (3rd ed.). Palgrave Macmillan.

Häkämies, S., Hirvonen, J., Jokisalo, J., & Knuuti, A. (2015). *Heat pumps in energy and cost efficient nearly zero energy buildings in Finland*. VTT Technology. https://gnf.fi/wp-content/uploads/2016/04/Report-Heat-pumps-in-energy-and-cost-efficient-nearly-zero-energy-buildings-in-Finland-_FINAL.pdf

Hardie, M., & Newell, G. (2011). Factors influencing technical innovation in construction SMEs: An Australian perspective. *Engineering, Construction and Architectural Management, 18*(6), 618–636. https://doi.org/10.1108/09699981111180926

Harty, C. (2008). Implementing innovation in construction: contexts, relative boundedness and actor–network theory. *Construction Management and Economics, 26*(10), 1029–1041.

Heaton, K. M., Skok, W., & Kovela, S. (2016). Learning lessons from software implementation projects: An exploratory study: Learning lessons from software projects. *Knowledge and Process Management, 23*(4), 293–306. https://doi.org/10.1002/kpm.1525

Home Innovation Research Labs. (2012). *Migliaccio*. https://www.homeinnovation.com/~/media/Files/Reports/2012_NGBS_Cost_Comparison.pdf

Hu, M., & Esram, N. W. (2021). The status of embodied carbon in building practice and research in the United States: A systematic investigation. *Sustainability, 13*(23), 12961. https://doi.org/10.3390/su132312961

Hughes, D., Williams, T., & Ren, Z. (2012). Belassi. *Construction Innovation, 12*(3), 355–368. https://doi.org/10.1108/14714171211244613

Hwang, B.-G., & Tan, J. S. (2012). Green building project management: Obstacles and solutions for sustainable development. *Sustainable Development, 20*(5), 335–349. https://doi.org/10.1002/sd.492

Hwang, B.-G., Zhu, L., & Ming, J. T. T. (2017). Factors affecting productivity in green building construction projects: The case of Singapore. *Journal of Management in Engineering, 33*(3), 04016052. https://doi.org/10.1061/(ASCE)ME.1943-5479.0000499

IEA. (2021). *Heat pumps.* https://www.iea.org/reports/heat-pumps

Kirkham, R. J., Brandon, P. S., & Ferry, D. J. (2015). *Ferry and Brandon's cost planning of buildings* (9th ed.). Wiley-Blackwell.

Koushki, P. A., Al-Rashid, K., & Kartam, N. (2005). Delays and cost increases in the construction of private residential projects in Kuwait. *Construction Management and Economics, 23*(3), 285–294. https://doi.org/10.1080/0144619042000326710

Kubba, S. (2012). *Handbook of green building design and construction: LEED, BREEAM, and Green Globes.* B-H Elsevier.

Logan, D. (2021, March 11). *Global exports of U.S.-grade framing lumber.* NAHB. https://eyeonhousing.org/2021/03/global-exports-of-u-s-grade-framing-lumber/#_ftn1

Lowe, D., Emsley, M., & Harding, A. (2006). Predicting construction cost using multiple regression techniques. *Journal of Construction Engineering and Management, 132*(7), 8. https://doi.org/10.1061/(ASCE)0733-9364(2006)132:7(750)

Lowe, D. J., Emsley, M. W., & Harding, A. (2006). Relationships between total construction cost and project strategic, site related and building definition variable. *Journal of Financial Management of Property and Construction, 11*(3), 165–180. https://doi.org/10.1108/13664380680001087

Magretta, J. (2012). *Understanding Michael Porter: The essential guide to competition and strategy.* Harvard Business Review Press.

Mao, C., Xie, F., Hou, L., Wu, P., Wang, J., & Wang, X. (2016). Cost analysis for sustainable off-site construction based on a multiple-case study in China. *Habitat International, 57*, 215–222. https://doi.org/10.1016/j.habitatint.2016.08.002

Mapp, C., Nobe, M., & Dunbar, B. (2011). The cost of LEED—An analysis of the construction costs of LEED and non-LEED banks. *Journal of Sustainable Real Estate, 3*(1), 254–273. https://doi.org/10.1080/10835547.2011.12091824

Meng, X., & Brown, A. (2018). Innovation in construction firms of different sizes: Drivers and strategies. *Engineering, Construction and Architectural Management, 25*(9), 1210–1225. https://doi.org/10.1108/ECAM-04-2017-0067

Migliaccio, G. C., Guindani, M., D'Incognito, M., & Zhang, L. (2013). Empirical assessment of spatial prediction methods for location cost-adjustment factors. *Journal of Construction Engineering and Management, 139*(7), 858–869. https://doi.org/10.1061/(ASCE)CO.1943-7862.0000654

Morton, R., & Jaggar, D. (1995). *Design and the economics of building* (1st ed.). E & FN Spon.

Naoum, S. G. (2016). Factors influencing labor productivity on construction sites: A state-of-the-art literature review and a survey. *International Journal of Productivity and Performance Management, 65*(3), 401–421. https://doi.org/10.1108/IJPPM-03-2015-0045

NBI. (2020). *2020 getting to zero buildings list.* https://newbuildings.org/resource/2020-getting-to-zero-project-list/

Nixon, P., Harrington, M., & Parker, D. (2012). Leadership performance is significant to project success or failure: A critical analysis. *International Journal of Productivity and Performance Management, 61*(2), 204–216. https://doi.org/10.1108/17410401211194699

Obi, L. I., Arif, M., & Kulonda, D. J. (2017). Prioritizing cost management system considerations for Nigerian housing projects. *Journal of Financial Management of Property and Construction, 22*(2), 135–153. https://doi.org/10.1108/JFMPC-06-2016-0025

Ofori, G. (2000). *Challenges of construction industries in developing countries: Lessons from various countries, 5*, 15–17.

Osipova, E., & Eriksson, P. E. (2011). How procurement options influence risk management in construction projects. *Construction Management and Economics, 29*(11), 1149–1158. https://doi.org/10.1080/01446193.2011.639379

Quillen, A. (2022). *Construction costs hit highest spike in 50 years*. https://www.nbcdfw.com/news/local/construction-costs-hit-highest-spike-in-50-years/2891677/

Rahman, I. A., Memon, A. H., Karim, A. A., & Tarmizi, A. (2013). Significant factors causing cost overruns in large construction projects in Malaysia. *Journal of Applied Sciences, 13*(2), 286–293.

Raouf Ayman, M., & Al-Ghamdi Sami, G. (2019). Effectiveness of project delivery systems in executing green buildings. *Journal of Construction Engineering and Management, 145*(10), 03119005. https://doi.org/10.1061/(ASCE)CO.1943-7862.0001688

Russ, N. M., Hanid, M., & Ye, K. M. (2018). Literature review on green cost premium of sustainable building construction. *International Journal of Technology, 9*(8), 1715. https://doi.org/10.14716/ijtech.v9i8.2762

Sanni-Anibire, M. O., Mohamad Zin, R., & Olatunji, S. O. (2020). Causes of delay in the global construction industry: A meta analytical review. *International Journal of Construction Management, 22*(8), 1395–1407. https://doi.org/10.1080/15623599.2020.1716132

Sanvido, V., Grobler, F., Parfitt, K., Guvenis, M., & Coyle, M. (1992). Critical success factors for construction projects. *Journal of Construction Engineering and Management, 118*(1), 94–111. https://doi.org/10.1061/(ASCE)0733-9364(1992)118:1(94)

Shane, J. S., Molenaar, K. R., Anderson, S., & Schexnayder, C. (2009). Construction project cost escalation factors. *Journal of Management in Engineering, 25*(4), 221–229. https://doi.org/10.1061/(ASCE)0742-597X(2009)25:4(221)

Shen, W., Tang, W., Wang, S., Duffield, C. F., Hui, F. K. P., & You, R. (2017). Enhancing trust-based interface management in international engineering-procurement-construction projects. *Journal of Construction Engineering and Management, 143*(9), 04017061. https://doi.org/10.1061/(ASCE)CO.1943-7862.0001351

Slaughter, E. S. (1998). Models of construction innovation. *Journal of Construction Engineering and Management, 124*(3), 226–231.

Sousa, P., Tereso, A., Alves, A., & Gomes, L. (2018). Implementation of project management and lean production practices in a SME Portuguese innovation company. *Procedia Computer Science, 138*, 867–874. https://doi.org/10.1016/j.procs.2018.10.113

Steven Winter Associates, Inc. (2004). *GSA LEED Cost Study Final Report* (Contract No. GS–11P–99–MAD–0565, Order No. P–00–02–CY–0065). https://archive.epa.gov/greenbuilding/web/pdf/gsaleed.pdf

Stoy, C., Pollalis, S., & Schalcher, H.-R. (2008). Drivers for cost estimating in early design: Case study of residential construction. *Journal of Construction Engineering and Management, 134*(1), 32–39. https://doi.org/10.1061/(ASCE)0733-9364(2008)134:1(32)

Sun, C.-Y., Chen, Y.-G., Wang, R.-J., Lo, S.-C., Yau, J.-T., & Wu, Y.-W. (2019). Construction cost of green building certified residence: A case study in Taiwan. *Sustainability, 11*(8), 2195. https://doi.org/10.3390/su11082195

Supply Chain Dive. (2022). *Construction material prices soared nearly 20% in 2021: Report*. https://www.supplychaindive.com/news/construction-materials-prices-soared-2021/617219/

Sweeney, C. (2022). *Four common construction contracts you need to understand*. https://www.aiacontracts.org/articles/183501-four-common-construction-contracts-you-need-to-understand-

Tasaki, T., Shobatake, K., Nakajima, K., & Dalhammar, C. (2017). International survey of the costs of assessment for environmental product declarations. *Procedia CIRP, 61*, 727–731. https://doi.org/10.1016/j.procir.2016.11.158

The International EPD System. (n.d.). *What is an EPD?* The International EPD System. Retrieved June 21, 2021, from https://www.environdec.com/all-about-epds/the-epd

The U.S. Census Bureau. (2022). *Monthly construction spending, January 2022.* https://www.census.gov/construction/c30/pdf/release.pdf

Tipili, L. G., & Ojeba, P. O. (2014). Evaluating the effects of communication in construction project delivery in Nigeria. In *Proceedings of the Multi-Disciplinary Academic Conference on Sustainable Development.* Multi-disciplinary Academic Conference on Sustainable Development, Federal Polytechnic, Bauchi.

Toh, T.-C., Ting, C., Ali, K.-N., Aliagha, G.-U., & Munir, O. (2012). Critical cost factors of building construction projects in Malaysia. *Procedia – Social and Behavioral Sciences, 57,* 360–367. https://doi.org/10.1016/j.sbspro.2012.09.1198

Tumminia, G., Guarino, F., Longo, S., Ferraro, M., Cellura, M., & Antonucci, V. (2018). Life cycle energy performances and environmental impacts of a prefabricated building module. *Renewable and Sustainable Energy Reviews, 92,* 272–283. https://doi.org/10.1016/j.rser.2018.04.059

Uğur, L. O., & Leblebici, N. (2018). An examination of the LEED green building certification system in terms of construction costs. *Renewable and Sustainable Energy Reviews, 81,* 1476–1483. https://doi.org/10.1016/j.rser.2017.05.210

USGBC. (2021). *Building product disclosure and optimization—Environmental product declarations.* https://www.usgbc.org/credits/new-construction-core-and-shell-schools-new-construction-retail-new-construction-data-43

Vrchota, J., Řehoř, P., Maříková, M., & Pech, M. (2020). Critical success factors of the project management in relation to industry 4.0 for sustainability of projects. *Sustainability, 13*(1), 281. https://doi.org/10.3390/su13010281

Yap, J. B. H., Abdul-Rahman, H., & Wang, C. (2018). Preventive mitigation of overruns with project communication management and continuous learning: PLS-SEM approach. *Journal of Construction Engineering and Management, 144*(5), 04018025. https://doi.org/10.1061/(ASCE)CO.1943-7862.0001456

Yoders, J. (2021). *2021 3Q cost report: No end in sight to year of supply chain chaos and materials challenges.* Engineering News-Record. https://www.enr.com/articles/52544-2021-3q-cost-report-no-end-in-sight-to-year-of-supply-chain-chaos-and-construction-logistics-material-challenges

Zhang, X., Platten, A., & Shen, L. (2011). Green property development practice in China: Costs and barriers. *Building and Environment, 46*(11), 2153–2160. https://doi.org/10.1016/j.buildenv.2011.04.031

Zhao, L., Mbachu, J., & Liu, Z. (2020). Identifying significant cost-influencing factors for sustainable development in construction industry using structural equation modelling. *Mathematical Problems in Engineering, 2020,* 1–16. https://doi.org/10.1155/2020/4810136

Zhao, L., Wang, B., Mbachu, J., & Liu, Z. (2019). New Zealand building project cost and its influential factors: A structural equation modelling approach. *Advances in Civil Engineering, 2019,* 1–15. https://doi.org/10.1155/2019/1362730

Zorich, M., & Flipsen, D. (2018). *Improving construction industry efficiency with IPD.* https://www.csemag.com/articles/improving-construction-industry-efficiency-with-ipd/

Zulch, B. (2014). Communication: The foundation of project management. *Procedia Technology, 16,* 1000–1009. https://doi.org/10.1016/j.protcy.2014.10.054

Part II

What actual data tells us

4   Data collection and description

ABSTRACT

Following the conclusion made in Chapter 3, due to the limitations of previous research findings and a lack of empirical data on the sustainable building construction cost (SBCC), the author decided to collect more data from reliable sources to create an SBCC database specific to this book. Chapter 4 first explains the data sources and collection procedures, followed by descriptive statistics for each data type.

4.1 SBCC COMPOSITION AND DATA SOURCES

The sustainable building construction cost (SBCC) database was created with three types of data: construction cost data, project team characteristic data, and project characteristic data. The cost data was obtained first, followed by the project and project team characteristic data. The main resources for the construction cost data were United States General Service Administration (GSA) records and the GSA website (for the project budget). The data for project characteristics were extracted from various documents found online, including project websites, project statements, and case studies conducted by GSA or other researchers. The project team characteristic data were also drawn from numerous sources, such as project reports, commissioned case studies, published journal papers, and phone interviews with project team members conducted by the author. An example is the Wayne Aspinall Federal Building and U.S. Courthouse, the GSA's first historic net-zero renovation project, certified as LEED Platinum. The National Renewable Energy Laboratory, together with GSA, produced a detailed 37-page case study report documenting the design and construction process of how the project achieved this high goal. Project team members were interviewed to draw on lessons learned and suggestions for further processes. Quotes from team members' interviews were extracted and included in the SBCC database.

4.2 DATA COLLECTION FROM GSA

GSA was established by President Harry Truman on July 1, 1949, to streamline the administrative work of the federal government. The mission of GSA has evolved from disposing of war surplus goods to currently providing stewardship of the way the government uses and manages real estate properties, acquisition services, and technology (GSA, n.d.-c). GSA owns and leases over 35 million square meter of space in 9,600 buildings in more than 2,200 communities nationwide. It is the largest public commercial real estate owner in the United States. In addition to office buildings, GSA properties include land ports of entry, courthouses, laboratories, post offices, and data processing centers.

Sustainability has been an evolving theme for GSA, which started with energy efficiency initiatives as a result of the oil embargo in the early 1970s. Following the first oil crisis, the focus of sustainability shifted from energy efficiency to overall sustainability in the early 1990s. In 2001, GSA was the first federal agency to join the U.S. Green Building Council (USGBC) and continues to play an active role to this day (GSA, 2008). As a result of a 2006 GSA evaluation of sustainable building rating systems, the administrator concluded that LEED is still the most credible rating system available to meet GSA's needs. Since then, GSA has increased its minimum requirement for new construction and substantial renovations of federally owned facilities to meet the LEED Certified level (*LEED Building Information*, n.d.). To date, GSA is the organization owning the most LEED-certified buildings. It is logical to choose GSA-owned LEED certified projects as a representation of sustainable building since studying projects from a single public agency can reduce the noise associated with clients' mindsets, financial capabilities, and other factors that often make studying the construction cost of different projects difficult.

According to the Freedom of Information Act (FOIA), on March 15, 2021, the author requested information on the construction costs of all LEED-certified government buildings. Through email correspondence with GSA, the researcher further clarified the FOIA request on March 25, March 29, and April 5. GSA agreed to provide a report of building-specific construction costs for all **175** GSA-owned LEED certified buildings that were on record as of March 31, 2021. On May 12, 2021, GSA provided the report with the agreed-upon information.

As illustrated in Figure 4.1, among the 175 buildings provided by GSA, 17 projects do not have cost information and were thus excluded, leaving 158 projects. To verify the LEED certification level of the case projects, the author cross-referenced the 158 projects with the USGBC online project directory; 45 projects are registered with USGBC but have not received final certification, thus they were excluded from the database. The cross-reference resulted in 113 projects at the end.

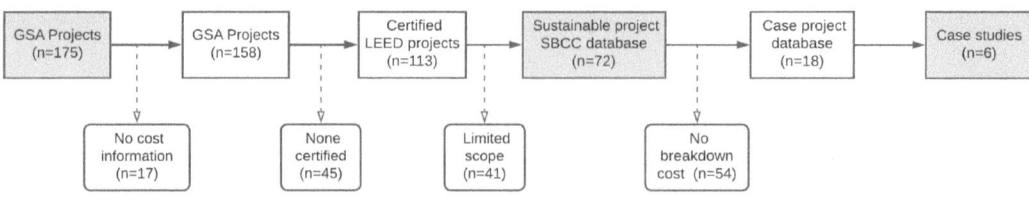

Figure 4.1
Data collection flow

After a preliminary examination of the 113 projects' costs, 41 projects were excluded due to a limited scope of work, and the remaining 72 projects were used to create the SBCC database. Accordingly, the project characteristic data and project team characteristic data were collected for the 72 projects to create a complete SBCC database and were then used to study the relation between the level of sustainability and the construction cost using a regression model. The SBCC database were also used for structural equation model (SEM) analysis to study the causal relations between project and project team characteristics and the construction cost.

The construction cost data provided by GSA do not include a detailed cost breakdown for major building components and systems. Using the GSA website, the author was able to find and download the budgeted cost breakdown of 18 projects, per major work performed for each major building system (e.g., mechanical system, structural system).

4.3 CONSTRUCTION COST DATABASE DESCRIPTIVE STATISTICS

4.3.1 Project cost range

The SBCC database created for this research indicates the construction cost of 37% of the buildings (first bar) is less than $538/m² ($50/ft²) as shown in Figure 4.2a. This low cost is related to the limited scope of work in the projects. Many of the renovation projects did not involve major building system upgrades or renovations; rather, they mainly included cosmetic repairs, such as repainting and minor repairs. Only a few projects in this category had building system upgrades (e.g., fire alarm system). This limited scope of work would not sufficiently reflect the technical complexity of sustainable buildings; therefore, those projects were excluded from further analysis.

After excluding the limited works, as illustrated in Figure 4.4b, overall, 84% of the projects fall under $1,122–$8,888/m² ($101–$800/ft²). Among the renovation projects included, the projects had at least four primary building systems completely renovated or replaced. In many of the renovation projects, the only system left untouched was the primary structure, which includes the foundation,

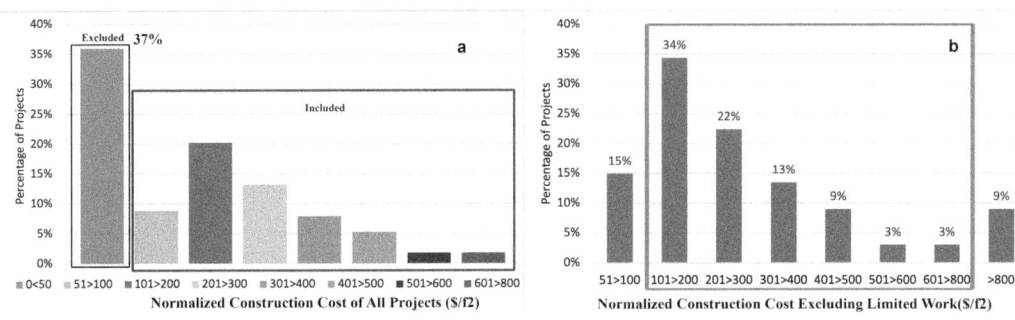

Figure 4.2
Construction cost distribution of projects in the SBCC database: (a) normalized construction cost of projects, and (b) normalized construction cost excluding limited scope of work projects

columns, and floors. The primary structure construction cost is less than 20% of the total construction cost of new commercial buildings.

Based on recent data (2021) from Cumming Insights Construction Market Analysis, the construction cost of a new mid-rise office building in the United States is around $6,244/m^2 ($562/ft^2); for high-rise buildings, the cost is $7,333/m^2 ($660/ft^2). If the cost of the primary structure is discounted (−20%), then the average construction cost of major renovation projects is $5,000/m^2 ($450/ft^2) and 5,866/m^2 ($528/ft^2) for mid-rise and high-rise offices, respectively. Therefore, according to this data, the GSA LEED buildings' construction cost is comparable to the conventional commercial office buildings' cost.

Within the range of $1,122–$8,888/m^2, the largest percentage (34%) of normalized costs is $1,122–$2,222/m^2, the second largest group (22%) is $2,223–$3,333/m^2, and the third largest group (13%) is $3,344–$4,444/m^2. A minimum of one project per group was selected as case studies for further investigation in Chapter 6. Exceptionally high-cost buildings (>$8,888/m^2) account for 9%; after reviewing the information of these projects, it was determined they were specific and did not represent a general condition of GSA projects, let alone normal sustainable buildings. For instance, four port of entry building complexes cost more than $10,000/m^2. Unlike regular administrative buildings, port of entry buildings have higher security requirements and other specific programs and space requirements related to port of entry functions (e.g., laboratory). These programmatic requirements are not applicable to other normal administrative buildings, thus the author did not further explore these projects.

4.3.2 Cost variance per geographic location

Regarding the cost difference per geographic location, Figure 4.3a indicates that location 7 has the highest average construction cost and largest cost variance, mainly due to the high construction cost in several major cities in California, such as San Francisco and Los Angeles. Location 2 has the second highest average construction cost, which is related to the high construction cost in major metropolitan cities in this region, such as Washington, D.C. Location 6 has the lowest average construction cost, followed by location 5. Two geographic locations – locations 1 and 2 – emerged with noteworthy results.

Location 1	Oregon, Washington	Location 5	Illinois, Indiana, Iowa, Minnesota, Missouri, Ohio
Location 2	District of Columbia, Maryland, Virginia	Location 6	Colorado, Montana, Utah, Wyoming
Location 3	Connecticut, Massachusetts, New Jersey, New York, Pennsylvania	Location 7	Arizona, California
Location 4	Texas	Location 8	Alabama, Florida, Georgia, Louisiana

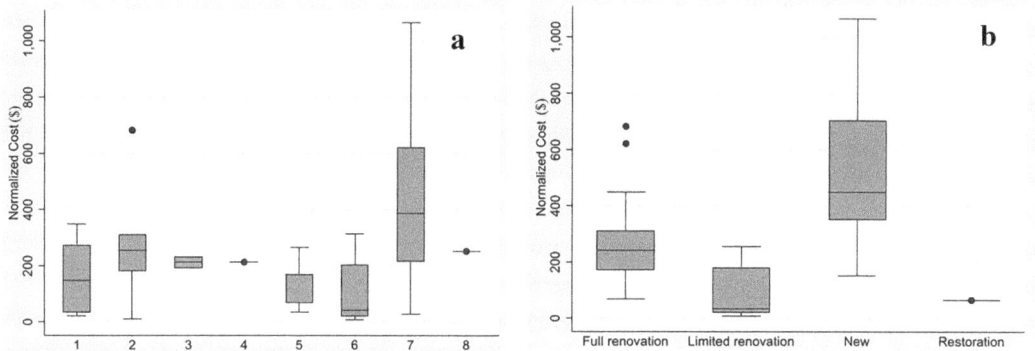

Figure 4.3
Cost variance studies (a) Cost variance per location. (b) Cost variance per construction type

As illustrated in Figure 4.3a, location 1 has a symmetrically distributed box plot diagram, which may indicate that the specific market (public projects) in this region is relatively mature compared to other regions. Another possible reason for a normal distribution is that the projects in location 1 are similar, in terms of type, complexity, and scope.

Although location 2 has the second highest average construction cost, the average price is still comparable to that of conventional buildings ($2,500/m²). In addition, the projects in location 2 have the least cost variance across building type, function, size, and age. This finding deviates from the common perception of expensive construction with a higher cost variance in the Washington, D.C. metropolitan region.

Both locations 1 and 2 have mean construction costs that are comparable to conventional buildings, or even lower than similar projects. For instance, in Washington, D.C., the average construction cost for a new convention mid-rise office building is $6,244/m²; for major renovations, it is $5,000/m². In the SBCC database, the average construction cost of projects in Washington, D.C. is $2,000/m², and all projects are major renovations, that is lower than conventional major renovation projects. After considering the annual inflation rate, GSA LEED buildings' construction cost is still comparable to that of conventional buildings in locations 1 and 2. A potential explanation for the competitive and even lower construction cost of these sustainable buildings is that the two regions may have highly skilled project teams who can execute complicated sustainable projects with lower cost. In addition, a relatively mature supply chain in these regions can provide reasonable prices for the building materials. The explanatory factors are further examined in the following chapter.

4.3.3 Cost variance per scope of work

Expectedly, new construction projects have a higher construction cost than renovated buildings, for both limited and full renovations. Figure 4.3b shows the average construction cost of a new GSA LEED building is $4,722/m², while fully renovated buildings have an average cost of $2,266/m². Two significant findings were discovered. *First*, new construction has the largest cost variance among projects, with the higher cost buildings (the portion above the median line) varying more than the lower cost buildings (the portion below the median line). *Second*, the box plot of fully renovated projects has a symmetrical configuration distribution with a couple of extreme outliers. This symmetrical distribution is different from the typical positively skewed distribution that represents fewer high-cost projects.

Typically, renovation projects carry more uncertainty than new construction due to incomplete information of the existing building's condition. Further, renovation projects are often seen as challenging because of site and technology constraints. Uncertainty and constraints can lead to a cost overrun and large cost variance among projects; consequently, a positively skewed distribution is often observed in renovation projects. However, results shown in Figure 4.3b suggest that **uncertainty and risk can be well managed** in renovation projects, and common practices can be derived from successful GSA LEED renovation projects. These findings provide clues for the selection of case projects for in-depth research in Chapter 6.

4.3.4 Cost variance per building type

Figure 4.4 illustrates a cost comparison among different building types. Ports of entry have the highest average construction cost and courthouses have the largest cost variance. Compared to other federal buildings, ports of entry have more specialty functions and often have a customized high-profile design that embodies the symbolic meaning of entry to the country. This high-profile design is often reflected in specialty construction and related higher costs, such as a double skin façade or a large overhanging PV panel-covered roof. These functions contribute to additional costs but not necessarily to the sustainability of the building. In addition, expenses for the interior furnishing are also ignored as a large contributor to the total construction cost. For example, the U.S. Land Port of Entry located in Columbus, New Mexico, had a total project cost of $96

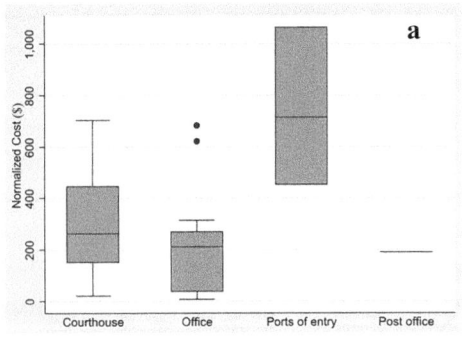
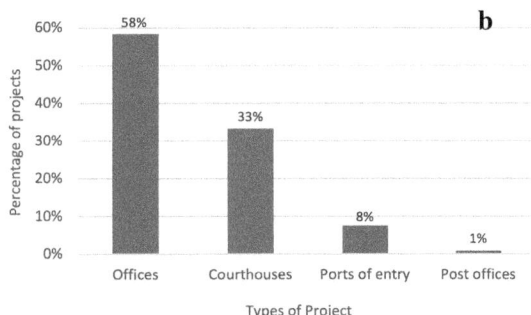

Figure 4.4
Cost variance studies (a) Cost variance per building type. (b) Building types included in the study

million (GSA, n.d.-d). The construction cost for site preparation, landscaping, and building construction was around $67 million (71%) (AIA, n.d.), while the cost for furnishing (e.g., furniture and interior decoration) was about $28 million (29%). Although the use of furniture and interior decoration can impact occupant health, it usually does not directly affect a building's energy efficiency.

4.3.5 Summary

When comparing the GSA LEED buildings to conventional (non-LEED) buildings, the average construction cost of a new GSA LEED building is lower than that of a conventional (mid-rise, non-LEED) office building: $4,722/m² versus $6,244/m² (Gerardi, 2021). The average construction cost of a GSA LEED renovation project is also lower than that of a conventional building: $2,266/m² versus $5,000/m².

Besides differences in spatial and program requirements and financial resources, a noticeable difference between GSA and non-GSA buildings is the **project team**. GSA projects, like other federally contracted projects, have screening requirements for the design and construction teams for security reasons. It is a lengthy process for firms and teams to obtain certification and clearance to work on GSA projects. Consequently, certified GSA contractors often receive repetitive opportunities to work on GSA projects, which helps design teams and contractors build up their experience and skills. Further, teams can develop better collaboration after working on several projects together. More experienced and skilled teams working together repetitively can contribute to controlling cost and budget overruns, consequently reducing the total construction cost.

4.4 PROJECT TEAM CHARACTERISTIC DATA COLLECTION

As explained in Chapter 3, seven indicators measuring the influence of project team characteristics are derived from the meta-analysis: *procurement & delivery methods, collaboration & communication* (Iyer & Jha, 2005, Meng et al., 2011), *skills & experience,* and *innovation* (Ozorhon & Oral, 2017). Table 4.1 demonstrates the data sources and types, followed by detailed explanations for each dimension.

4.4.1 Procurement and delivery methods

When Congress enacted the American Recovery and Reinvestment Act (ARRA) in 2009 in response to the Great Recession of 2008, the primary objective of this federal status was to save existing jobs and create new ones. GSA received $5.5 billion in funds to convert federal buildings into high-performance green buildings (Burrell, 2016), and another $750 million was allocated to federal building and courthouse renovations (Chang et al., 2014). Prior to ARRA, GSA was on the

Table 4.1 Data categories, resources, and values

Data	Data resource	Description	Data type
Procurement & delivery methods	Online project website, statements, records	1 = Best value 2 = Low bid 3 = Sole source selection 4 = Qualifications-based	Quality
Skills & experience	Online ENR ranking	1 = No skills 2 = Few skills 3 = Moderate skills 4 = Specialists 5 = Leading experts	Quality
Collaboration & communication	(i) Interviews (ii) publicly accessible records (iii) number of communication channels	1 = Very poor 2 = Below average 3 = Average 4 = Above average 5 = Excellent	Quality & quantity
Innovation	(i) Client satisfaction (ii) number of innovation items (iii) LEED points (iv) project cost and schedule	1 = Very poor 2 = Below average 3 = Average 4 = Above average 5 = Excellent	Quality & quantity

verge of disposing several old buildings, including the Wayne N. Aspinall Federal Building and U.S. Courthouse (see details in Chapter 6). When the funds became available mandating a high-performance goal, GSA was able to align renovation needs with the energy efficiency goal and implement many projects in a short period, between **2010 and 2019**. In GSA's FY 2012 Sustainability Plan, it stated:

> GSA remains committed to leading the federal government towards a more sustainable future. GSA recognizes the important role it plays in facilitating sustainability goal achievement for other federal agencies. GSA will continue to identify and make available innovative, cost effective, and sustainable solutions for federal agencies. GSA will also continue to examine ways in which it can use its sustainability efforts to improve transparency of its operations; reduce costs and eliminate waste; improve accountability and agency performance management; and fully integrate sustainability into its day-to-day operations.
>
> (GSA, 2012)

GSA's determination and the constrained budget and schedule prompted the need for a capable team that could deliver a high-performance building within the budget. Finding a skilled team relies on the type of procurement method. Chapter 3 described four commonly used procurement methods – *qualification-based selection*, *best value selection*, *low-bid selection*, and *sole source selection*—and four common project delivery methods – *design-bid-build (DBB), design-build (DB), construction management (CM), and integrated project delivery (IPD)*.

In a DBB project, for the design service, GSA normally uses a qualification-based selection process under the Brooks Act. It is a two-step process where technical submissions of qualification from architectural/engineering (A/E) firms are received, and those with the strongest technical proposals are interviewed to develop a list of finalists. The firms are ranked based on their technical qualifications, and negotiations are conducted with the top-rated firm for the contract award (GSA, n.d.-a). For construction services, major construction contracts are selected using the source selection method, which is a combination of technical qualifications and a price proposal. Once the proposals are received, they are evaluated technically and then assessed in terms of price, with the final selection based on best value.

When GSA issued the first edition of its cutting-edge *Design Excellence Policies and Procedures* in 1994, the federal government rarely used the DB delivery method. Instead, most GSA construction projects used the DBB model, and the government procured the services of an A/E firm pursuant to the Brooks Act and its implementing regulations. Since that original publication, the use of DB in the federal government has expanded considerably, especially at GSA. One reason is that Congress passed legislation to provide special rules and procedures for acquiring DB services. Similarly, the Federal Acquisition Regulation (FAR) was updated to account for the new procedures. In 2016, GSA assembled a team of in-house experts in the field of architecture and construction, as well as various industry partners, to add a new chapter to *Design Excellence Policies and Procedures*, Chapter 9, that focuses on the DB delivery method (GSA, n.d.-b). The purpose of this added chapter is to streamline the use of the DB delivery method by providing cohesive and uniform policies. This effort plays an important role in how the DB delivery method is used in many high-performance building projects that are included in this book.

4.4.2 Skills and experience

In this book, skills refer to the project design team's (excludes client) collective skills, which are measured based on the company's (firm's) ranking in the two most recognized ranking systems in the building and construction sector. The first ranking is the *Top 100 Green Buildings Design Firms* by Engineering News-Record (ENR) from 2008 to 2020. ENR is widely regarded as one of the construction industry's most reliable publications in the United States. Companies are ranked according to revenue for construction or design services generation in the previous year from projects that have been registered with or certified by a third-party organization that sets the standard for measuring a building's or facility's environmental impact, energy efficiency, or carbon footprint. Such groups include the U.S. Green Building Council (USGBC) and Green Building Initiative. The second ranking is *Architect 50: Top 50 Firms in Sustainability*, published by

Architect, the official journal of the American Institute of Architects (AIA), from 2012 to 2019. The sustainability score is derived from five categories: (i) the 2030 commitment, (ii) the energy and water conservation target achieved by projects, (iii) the number of LEED certified employees, (iv) the number of certified green buildings (LEED, Living Building Challenge, Green Globe, Net Zero, Green Guide for Health Care, Energy Star, Passive House, and other leading certifications), and (v) the green project that best demonstrated a firm's commitment to sustainability. The five categories are weighted differently, at 18%, 18%, 6%, 20%, and 38%, respectively.

These two rankings have some overlap. For example, HOK, as a large design firm, is ranked fourth on the 2020 ENR 2020 Top 100 Green Buildings Design Firms list and tenth on the 2019 Architect 50: Top 50 Firms in Sustainability. ZGF Architects LLP is listed as fourteenth on the ENR list and twenty-first on the *Architect* list. In these cases, this book used the average of the two rankings as an overall ranking to measure the design team's skill.

The design team's experience is measured by its company's overall construction/project experience in sustainable building. The experience level is also based on the ENR ranking. As explained in the last section, many projects used the DB project delivery method, which entails a combined designer and contractor team; therefore, two ENR rankings were to provide a comprehensive score. The first ranking is ENR's Top 100 Green Building Contractors, from 2007 to 2020. The ranking is based on construction revenue in the previous year in ($) millions. Lists from different years were combined, and the final ranking was based on an average ranking of those years. The second ranking is ENR's Top 500 Design Firms, from 2003 to 2020, and this list includes different firm types, such as architect, engineer, engineer-contractor, architect-engineer, engineer-architect, environmental engineer, geotechnical engineer, and landscape architect.

These two rankings have overlapping as well. For example, AECOM, a large design and construction firm, is ranked fourth in 2020 Top 100 Green Building Contractors and second in 2020 Top 500 Design Firms. In this case, this book used the average of the two rankings as an overall ranking to measure the design team's skill.

The final ranking of the project team is the average of the two ranking scores; for example, if one project uses AECOM as the design firm (ranked second) and Hensel Phelps as the contractor (ranked twelfth), then the final experience score for the project team is seven.

4.4.3 Collaboration and communication

Good collaboration and communication between architects, engineers, and quantity surveyors is critical to controlling the construction cost (Elinwa & Buba, 1993). The success of a project is impacted by how efficiently team members can communicate with each other. Conversely, overcommunication can weaken a team's productivity and ultimately lead to burnout. Communication overload can occur if there is no proper communication protocol established within the project team (Segerstedt & Olofsson, 2010).

In this book, team communication is defined as a sharing of information between two or more individuals or groups to reach a common understanding. It is measured by three items: (i) focused interviews conducted by the researchers with project teams; (ii) self-reporting and evaluation, including publicly accessible records; and (iii) communication channels using the formula presented in *A Guide to the Project Management Body of Knowledge*.

During the focused interviews, three questions related to team communication were asked using a Likert scale of 1–5, as listed in Table 4.2.

For the self-report and evaluation, a benefit of choosing publicly funded buildings, such as GSA LEED buildings, is that GSA often conducts assessments to evaluate project performance, and these reports are in the public domain. For instance, the Wayne Aspinall Federal Building and U.S. Courthouse project was GSA's first net-zero energy historical preservation project, with key lessons learned from the project team including conducting "regular meetings with open communication" (Chang et al., 2014). Because this project had the dual goals of a net-zero energy retrofit and historical preservation, there was a unique set of risks faced by the integrated DB team; for instance, conflict between the aesthetic impact of the originally proposed large solar panel overhang and historic preservation guidelines. Considering these uncommon risks, early collaboration was essential, and the team emphasized strong relationships and an open-minded approach to achieve a collaborative culture (Cheng, 2015). Another building included in the case study report is Federal Center South Building 1202, located in Seattle, Washington. The DB team included ZGF Architects LLP and Sellen Construction. It was also a renovation project and achieved LEED Gold

Table 4.2 Focused interview questions about team communication

Items	Description	Rating
1	Communication process (timely and organized)	1 = Very poor 2 = Below average 3 = Average 4 = Above average 5 = Excellent
2	Communication effectiveness (concise and clear)	1 = Very poor 2 = Below average 3 = Average 4 = Above average 5 = Excellent
3	Communication style (open and honest)	1 = Never 2 = Seldom 3 = Sometimes 4 = Often 5 = Excellent

certification. Team members stated, "Sellen construction allowed the design team [to have] a little more rope than a lot of other contractors do, allowing [the] designer to explore ideas and have enough time to let good ideas really come to the top." Because "there was a high level of trust already established…. We didn't have to build trust. We could go beyond that and just perform" (Cheng, 2015).

The reason for choosing communication channels as a measure is because they show how information is shared and how it flows within the team. Communication channels are dependent on the number of key independent players in the project team and reflect the procurement type to a certain extent. For example, in a DBB project, each consultant that is hired and reports to the client is considered an independent player. However, in DB projects, all consultants fall under the design team, which can be combined as one firm, or one player. The number of channels is calculated using Equation 4.1, adopted from *A Guide to the Project Management Body of Knowledge* (PMBOK, 6th edition).

$$\text{Communication channels} = N * (N - 1) / 2, \qquad \text{(Equation 4.1)}$$

where N is the number of primary consultant firms hired; this information was extracted from the project website and project statement.

The final score for collaboration and communication is the aggregated score of the three measures described above and listed in Table 4.1.

4.4.4 Innovation

4.4.4.1 Innovation input

The input of innovation in the construction industry has been identified in various formats, such as *financial investment, human resources allocation, internal knowledge generation, knowledge transfer,* and *consultancy* (Ozorhon & Oral, 2017). Financial investment mainly refers to the money invested in creating or adopting new knowledge, research and development (R&D), and organizational practices. The construction industry has a reputation of being less innovative, according to its limited commitment to R&D expenditures; for instance, few design and construction firms use financial resources to design new tools or specialized equipment (Tatum, 1987). Similar to financial investment, innovation success can be ensured by expanding the innovation team; effective composition of innovation teams has been identified as a critical factor (Warszawski, 1996).

Knowledge generation refers to a company learning from its own experiences (internal knowledge generation) as well as from external sources (external knowledge generation). Knowledge can be transferred from one project to another within the company or from outside. Problems arising during construction (on-site) can foster innovation, and the roles of the project management team and site personnel are significant if they can transfer an innovative solution from one site (project) to others. However, knowledge and innovation are not always transferable between projects. Most construction and design firms do not develop new technology; rather, they may introduce technology from outside the construction and design industry (Bossink, 2004). For example, SketchUp, a digital design technology and tool prevalently used by architects, was created by Brad Schell, a structural engineer, and Joe Esch, a computer science engineer (SketchUp, n.d.).

The last format of innovation input in the building and construction industry is consultancy. To foster innovative competencies, consultants establish working relationships with the client, partnering with technology suppliers to absorb new technologies and formalize strategic plans (Ozorhon & Oral, 2017). When the knowledge cannot be obtained within the company or easily introduced from outside, then working with a specialized consultant to form a project-specific solution can be an effective way for innovation. For example, for the Edith Green-Wendell Wyatt Federal Building renovation project, the original DB team lacked experience and expertise in radiant cooling system design and construction; therefore, GSA requested that the team hire an additional consultant who was an expert in the area.

The above-mentioned five inputs of innovation represent different types of human and monetary resources used to facilitate innovation. The input factors lead to different outputs, which are explained below.

4.4.4.2 Innovation output measures

Although many measures have been used extensively to quantify innovation – such as the number of patents issued by a firm, key performance indicators (Banu, 2018), and R&D output – they are not sufficient to measure innovation in the construction industry since the industry is project-based and the organizational context of construction innovation differs significantly from many manufacturing innovations (Ozorhon & Oral, 2017). This book focuses on project-specific factors of innovation in sustainable building projects. Therefore, in addition to the commonly used measures applicable to all industries, the innovation output from sustainable projects can be measured using metrics associated with input, such as a decrease in project duration (Ozorhon et al., 2014) or project cost (Ozorhon, 2013) and an increase in productivity (Bröchner & Olofsson, 2012) or client satisfaction (Tang et al., 2003). Among these potential measures, the author selected four specific metrics based on the availability of the data: (i) client satisfaction, (ii) the number of specific innovation items (e.g., radiant cooling system was counted as one item) (iii) the credits received under the LEED Innovation category, and (iv) project cost and schedule influence.

For client satisfaction, information were extracted from project reports and focused interviews with key project team members. During the focused interviews, three questions were asked of the project members, as listed in Table 4.3.

Regarding the number of specific innovation items, the data were also extracted from project reports. A report typically has a designated category to describe the innovation of the project. For example, the radiant cooling panel is identified as a major innovation item in the Edith Green-Wendell Wyatt Federal Building project report.

Table 4.3 Focused interview questions on innovation

Items	Description	Rating
1	Attitude toward innovation (welcoming and enthusiastic)	1 = Very poor 2 = Below average 3 = Average 4 = Above average 5 = Excellent
2	Effort toward innovation	1 = Very poor 2 = Below average 3 = Average 4 = Above average 5 = Excellent
3	Outcome of innovation (sustainability and/or cost saving)	1 = Unknown 2 = None 3 = A little 4 = Some 5 = A lot

For credits received under the LEED Innovation category, a total of six credits are available regardless of the different rating systems (a sample LEED scorecard can be found in Chapter 6). The intent of the Innovation credit category is to encourage projects to achieve an exceptional or innovative performance and covers innovation in design intent, design requirements, and design implementation. Innovation in design intent awards design teams and projects for performance that exceeds requirements set by LEED, or innovative performance in sustainable building categories not addressed by the LEED rating system. Innovation in design requirements asks design teams to identify the intent of the proposed innovation credit (in writing) and demonstrate code compliance. Various innovations have been submitted, such as innovative sustainable waste management and design for flexibility. Innovation in design implementation requires proof of implemented innovation strategies.

Lastly, project cost and schedule impact indicators, such as whether the project was delayed or experienced cost overrun, are used to measure innovation effectiveness.

The final score for innovation is the aggregated score of the four metrics described above and listed in Table 4.1.

4.5 PROJECT CHARACTERISTIC DATA DESCRIPTIVE STATISTICS

4.5.1 Building type

GSA buildings include six building types: offices, land ports of entry, courthouses, laboratories, data processing centers, and post offices. ARRA funded a total of 261 GSA projects in 50 states, and among the $4.5 billion allocated to high-performance green buildings, $750 million was allocated to federal building and courthouse renovations and $300 million to land port of entry renovation and construction (AIA Minnesota Convention, 2015). However, the information acquired by the author through FOIA only included 175 projects under ARRA, all which were federal offices, land ports of entry, and courthouses. As demonstrated in Figure 4.4b, the projects included in this study are offices (58%), courthouses (33%), ports of entry (8%), and post offices (1%). Although the studied projects do not represent the full range of GSA-owned buildings, the analysis results from this dissertation research can be used to understand the conditions and status of GSA-owned LEED projects.

4.5.2 Building age
Thirty-five percent of the buildings were built in 2001–2020, 26% in 1961–1980. Twenty-one percent in 1921–1940, 7% in 1941–1960, 6% before 1920, and only 4% in 1981–2000.

4.5.3 Project location distribution
Figure 4.5a shows the geographic distribution of the projects. District of Columbia has the most projects ($n = 16$), followed by California ($n = 11$), Colorado ($n = 9$), and Washington ($n = 6$). There are eight states with only one project: Georgia, Iowa, Louisiana, Minnesota, Montana, New Hampshire, Wisconsin, and Wyoming. In general, the projects included in this study represent the project population of GSA buildings in terms of geographic location.

Regarding floor area, as illustrated in Figure 4.5b, District of Columbia has the largest floor area, followed by California and New York, which is mainly due to the large number of federal office buildings located in the District of Columbia.

4.5.4 Project level of sustainability achieved
Figure 4.6a illustrates the LEED certification breakdown of the studied projects: 41% of projects are Certified, 25% are Silver, 28% are Gold, and 6% are Platinum. Figure 4.6b shows that despite District of Columbia having the highest number of LEED buildings, most of these buildings are only Certified, while California leads in the level of sustainability with the most Platinum buildings ($n = 5$).

4.5.5 Construction type (scope of work)
Regarding the type of construction of the studied projects, the majority of projects (53%) had a limited renovation scope, 26% of projects were full renovations, and 20% were new construction. A fully renovated project is defined in this book as a project that has at least four primary building systems fully renovated, retrofitted, or upgraded. The primary building systems are mechanical systems (e.g., heating, cooling, ventilation), plumbing systems, electrical systems, building envelope systems, fire safety systems, and vertical transportation systems (e.g., elevators). In later case studies, projects with a limited scope

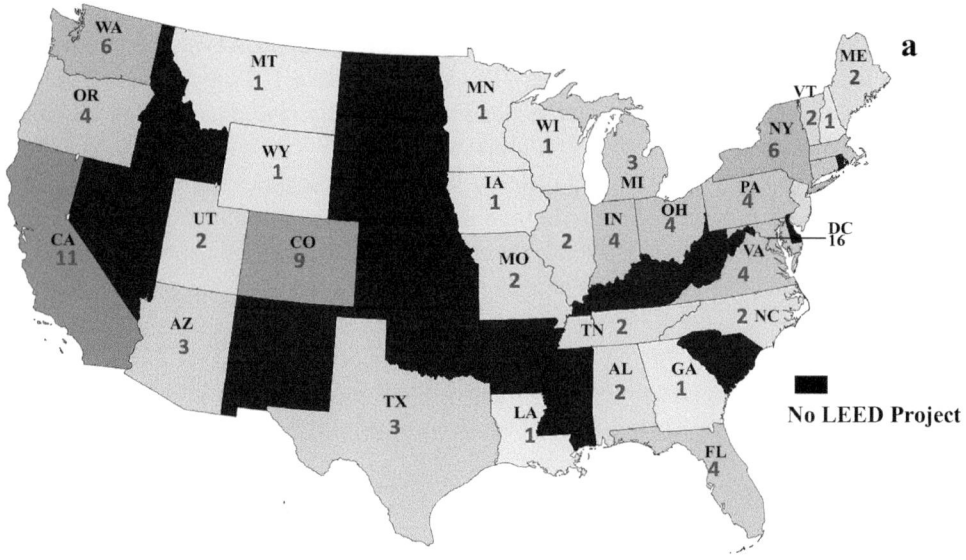

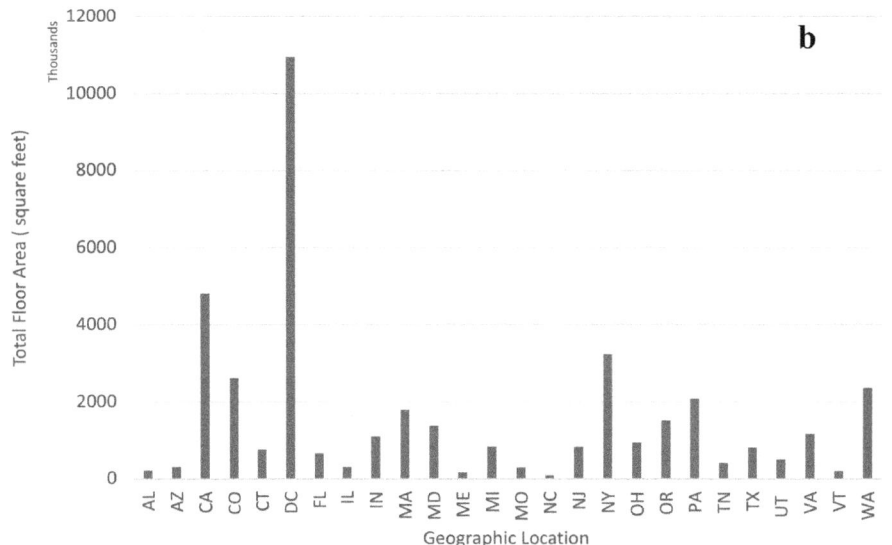

Figure 4.5
GSA LEED project distribution (a) GSA LEED project location map (created by author based on raw data from GSA).
(b) Studied project floor area per location

of work were excluded to gain a comprehensive and more accurate understanding of the construction cost factors.

Overall, the projects included in this book well represent GSA-owned facilities across different geographic locations at different LEED certification levels. The construction cost data were provided by the project owners, which is one of the most reliable data sources. Since there were robust documents and GSA-commissioned studies on lessons learned from LEED certified projects, data on

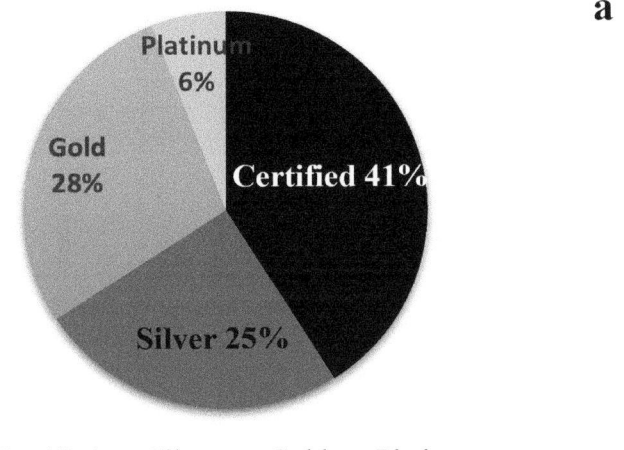

LEED Certification Level

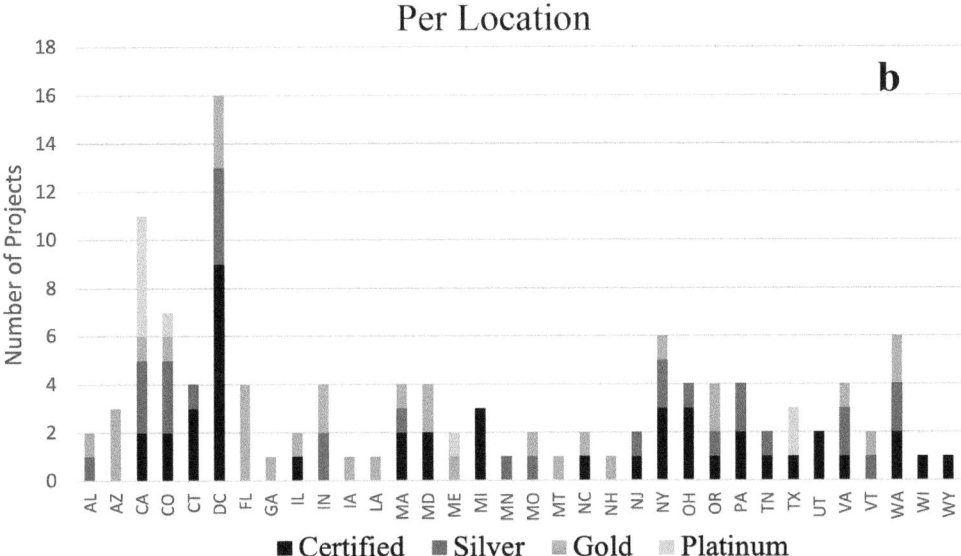

Per Location

Figure 4.6
LEED certification levels of studied projects: (a) certification level breakdown, and (b) certification geographic distribution

project team characteristics were successfully extracted, even at the level of direct quotes from team members.

4.6 CHAPTER SUMMARY

Chapter 4 first explains the data collection process and data sources. Then, a descriptive analysis of data and examination of the qualitative data are provided.

Overall, this chapter prepares the database for a statistical analysis and case studies, which will be explained in Chapters 5 and 6. One important point derived from the SBCC descriptive statistics is that the **SBCC is comparable to the conventional building** construction cost, if not lower.

REFERENCES

AIA. (n.d.). *U.S. Land Port of Entry, Columbus, New Mexico*. 2020 COTE Top Ten. Retrieved April 12, 2022, from https://www.aia.org/showcases/6280250-us-land-port-of-entry-columbus-new-mexico

AIA Minnesota Convention. (2015). *GSA high performance green building success stories*. http://www.aia-mn.org/wp-content/uploads/Ev30-GSA-Green-Building-Success-Stories.pdf

Banu, G. S. (2018). Measuring innovation using key performance indicators. *Procedia Manufacturing, 22*, 906–911. https://doi.org/10.1016/j.promfg.2018.03.128

Bossink, B. A. G. (2004). Managing drivers of innovation in construction networks. *Journal of Construction Engineering and Management, 130*(3), 337–345. https://doi.org/10.1061/(ASCE)0733-9364(2004)130:3(337)

Bröchner, J., & Olofsson, T. (2012). Construction productivity measures for innovation projects. *Journal of Construction Engineering and Management, 138*(5), 670–677. https://doi.org/10.1061/(ASCE)CO.1943-7862.0000481

Burrell, M. (2016, February 18). *GSA looks back at the American recovery and reinvestment act*. https://www.gsa.gov/blog/2016/02/18/gsa-looks-back-at-the-american-recovery-and-reinvestment-act

Chang, R., Hotchkiss, S., Pless, S., Sielcken, J., & Smith-Larney, C. (2014). *Aspinall Courthouse: GSA's historic preservation and net-zero renovation: Case study*. https://www.energy.gov/eere/femp/downloads/aspinall-courthouse-gsa-s-historic-preservation-and-net-zero-renovation

Cheng, R. (2015). *Integration at its finest: Success in high-performance building design and project delivery in the federal sector*. Research Report. Office of Federal High-Performance Green Buildings, U.S. General Services Administration.

Elinwa, A. U., & Buba, S. A. (1993). Construction cost factors in Nigeria. *Journal of Construction Engineering and Management, 119*(4), 698–713. https://doi.org/10.1061/(ASCE)0733-9364(1993)119:4(698)

Gerardi, J. (2021, June). *Commercial construction costs per square foot*. https://proest.com/construction/cost-estimates/commercial-costs-per-square-foot/

GSA. (n.d.-a). *Bidding on federal construction projects*. Retrieved April 12, 2022, from https://www.gsa.gov/real-estate/real-estate-services/for-businesses-seeking-opportunities/bidding-on-federal-construction-projects

GSA. (n.d.-b). *Design excellence: Policies and procedures*. Retrieved April 12, 2022, from https://www.gsa.gov/real-estate/design-and-construction/design-excellence/design-excellence-program/design-excellence-policies-and-procedures

GSA. (n.d.-c). *Our mission's evolution*. Retrieved April 6, 2022, from https://www.gsa.gov/about-us/mission-and-background/our-missions-evolution

GSA. (n.d.-d). *U.S. land port of entry at Columbus, NM*. Retrieved April 12, 2022, from https://www.gsa.gov/about-us/regions/welcome-to-the-greater-southwest-region-7/buildings-and-facilities/us-land-port-of-entry-at-columbus-nm

GSA. (2008). *Sustainability matters*. U.S. General Services Administration. https://www.gsa.gov/cdnstatic/Sustainability_Matters_508.pdf

GSA. (2012). *GSA FY 2012 sustainability plan*. https://www.gsa.gov/cdnstatic/AgencyManagementCommentsontheInspectorGeneralsAssessment.pdf

Iyer, K. C., & Jha, K. N. (2005). Factors affecting cost performance: Evidence from Indian construction projects. *International Journal of Project Management, 23*(4), 283–295. https://doi.org/10.1016/j.ijproman.2004.10.003

Meng, X., Sun, M., & Jones, M. (2011). Maturity model for supply chain relationships in construction. *Journal of Management in Engineering, 27*(2), 97–105. https://doi.org/10.1061/(ASCE)ME.1943-5479.0000035

Ozorhon, B. (2013). Analysis of construction innovation process at project level. *Journal of Management in Engineering, 29*(4), 455–463. https://doi.org/10.1061/(ASCE)ME.1943-5479.0000157

Ozorhon, B., Abbott, C., & Aouad, G. (2014). Integration and leadership as enablers of innovation in construction: Case study. *Journal of Management in Engineering, 30*(2), 256–263. https://doi.org/10.1061/(ASCE)ME.1943-5479.0000204

Ozorhon, B., & Oral, K. (2017). Drivers of innovation in construction projects. *Journal of Construction Engineering and Management, 143*(4), 04016118. https://doi.org/10.1061/(ASCE)CO.1943-7862.0001234

Segerstedt, A., & Olofsson, T. (2010). Supply chains in the construction industry. *Supply Chain Management: An International Journal, 15*(5), 347–353. https://doi.org/10.1108/13598541011068260

SketchUp. (n.d.). *Brad Schell Founder & Board Member, Sketchup (acquired by Google).* Retrieved April 19, 2022, from http://techstars.hypersites.com/site/page/pg5887-as691.html

Tang, S. L., Lu, M., & Chan, Y. L. (2003). Achieving client satisfaction for engineering consulting firms. *Journal of Management in Engineering, 19*(4), 166–172. https://doi.org/10.1061/(ASCE)0742-597X(2003)19:4(166)

Tatum, C. B. (1987). Process of innovation in construction firm. *Journal of Construction Engineering and Management, 113*(4), 648–663. https://doi.org/10.1061/(ASCE)0733-9364(1987)113:4(648)

U.S. General Services Administration. (n.d.). *LEED building information.* Retrieved June 12, 2021, from https://www.gsa.gov/real-estate/design-construction/design-excellence/sustainability/sustainable-design/leed-building-information

Warszawski, A. (1996). Strategic planning in construction companies. *Journal of Construction Engineering and Management, 122*(2), 133–140. https://doi.org/10.1061/(ASCE)0733-9364(1996)122:2(133)

5 Data analysis

ABSTRACT

This chapter presents the statistical analysis findings derived from the sustainable building construction cost (SBCC) database collected for this book. The quantitative and qualitative analysis and the hypothesis testing process are explained to help facilitate readers' understanding. The main objective of the data analysis is (a) to gain an understanding of whether there is a relation between the construction cost and level of sustainability (LOS), and (b) to determine a causal relation between the project characteristics and project team characteristics and the construction cost. A regression model and structural equation model (SEM) were two statistical tools used for analysis. Ten hypotheses were tested, followed by an explanation of the results, with the primary findings highlighted below.

HIGHLIGHTS

Technical complexity is not a significant factor to the final construction cost.

The **design cost** is as influential as the hard costs on the total project construction cost.

The **project team characteristics** are more influential than the project characteristics on the total project construction cost.

Project team skills and experience are the dominating variables.

NOMENCLATURE

EFA	Exploratory factor analysis
KMO	Kaiser-Meyer-Olkin
LEED	BD+C LEED for Building Design and Construction
LEED	O+M LEED for Existing Buildings Operations and Maintenance
LOS	Level of sustainability
MLE	Maximum likelihood estimation
SBCC	Sustainable building construction cost
SEM	Structural equation model

5.1 REGRESSION MODEL ANALYSIS

5.1.1 Relation between the level of sustainability and the construction cost

Two regression models were created to study the correlation between the normalized sustainable building construction cost (SBCC) and the level of sustainability (LOS) those buildings achieved. *Hypothesis 1a (H1a)* states there is no correlation between the construction cost and the level of sustainability. In this book, two methods were used to measure the LOS: the LEED certification level and the overall LEED score the projects received.

The first measure involved assigning the four LEED certification levels (Certified, Silver, Gold, and Platinum), where 1 represents Certified and 4 represents Platinum. The second measure used the normalized LEED score. Certain points are required to achieve each certification level, and the number of points required for a certain level varies depending on which LEED rating system and version are applied to the projects. For example, the current LEED v4.1 for Building Design and Construction (LEED BD+C) has a maximum of 110 points; to achieve the Certified level, at least 40 points are required. Silver requires 50–59 points, Gold requires 60–79 points, and Platinum requires 80 or more points. However, the third version of LEED BD+C, released in late 2005, only has a total of 69 available points; 26–32 points are required for Certified, 33–38 points for Silver, 39–51 points for Gold, and 52–69 points for Platinum. Meanwhile, LEED v4.1 O+M, Existing Buildings Operations and Maintenance, has a maximum of 100 points, but the certification levels are the same as LEED BD+C. The projects included in this study are certified under different rating systems, and/or under different versions. A summary of the different systems and points are listed in Table 5.1.

Because of the differences among different LEED versions, for a fair comparison, the LEED score was normalized by the percentage of points received for certification. For example, in the LEED BD+C (v2009) system, there is a maximum of 110 points. A Silver project that received 54 points is calculated as 54/110, equaling 0.49. In the LEED O+M system (for existing buildings), there is a total of 100 points. Therefore, a Silver project that received 54 credits is calculated as 54/100, equaling 0.59.

Using different measures, the author received different results. As shown in Table 5.2, when measured using the normalized LEED score, there is no statically significant correlation ($p > 0.05$) between the normalized SBCC and the LOS. Conversely, when measured using the LEED certification level, the data shows

Table 5.1 Certification levels and points of the rating systems

LEED system	Available points	Certified	Silver	Gold	Platinum
BD+C (v2009)	110	40–49	50–59	60–79	80+
O+M	100	40–49	50–59	60–79	80+
ID+C	110	40–49	50–59	60–79	80+
BD+C (v2.1)	69	26–32	33–38	39–51	52–69

Table 5.2 Regression analysis between the LOS and construction cost

Measure (LOS)	R-squared	p-value	Coefficients
LEED normalized score	0.066	0.00576	515.85
LEED certification level	0.194	**1.0512E−06**	106.71

Notes: Bold values represent the statistical significance. A result is considered statistically significant if the probability of it occurring by chance is less than a certain value, usually 0.05. This means that if the experiment were repeated many times, the result would only occur by chance 5% of the time or less.

a statistically significant correlation ($p < 0.05$) between the normalized SBCC and the level of certification (Certified, Silver, Gold, and Platinum). However, only 19.4% of the variance of the cost difference can be explained by the LOS.

Based on the results from the two regression models, the relation between the normalized SBCC and the LOS was *inconclusive*. Although there is a correlation between the two, the correlation coefficient is very low (R-squared = 0.194), which means only around 19.4% of the difference in the construction cost can be explained by the different LEED certified levels. **This is an indication that there are other parameters that critically influence the construction cost of LEED buildings, other than the LOS achieved**. To uncover these hidden influences, a deep dive into the construction cost data was conducted in the following section.

5.1.2 Construction cost per LOS

Because the results for H1a were inconclusive, to understand how the construction costs are reflected in different sustainability levels, then, the projects were examined according to their LOS (measure by LEED points). As illustrated in Figure 5.1, the circles represent LEED Certified, diamonds represent LEED Silver, *X*'s represent LEED Gold, and squares represent LEED Platinum. The *Y*-axis indicates the percentage of LEED points the projects obtained, where more points signify a higher reward, certification level, and LOS.

Figure 5.1 shows the average cost of a LEED building is $3,100/m² ($288/ft²). The average cost of a Certified building is $1,001/m² (~$93/ft²), Silver is $1,733/m² (~$161/ft²), Gold is $4,015/m² (~$373/ft²), and Platinum is $3,466/m² (~$322/ft²). The prices are the average values between 2003 and 2020. In general, there is **NO** clear indication that a higher SBCC is associated with a higher rating level and score. The sample data suggest that LEED Gold level projects have a higher average cost than that of LEED Platinum buildings, which indicates there are other factors moderating the cost of higher LEED level projects.

As illustrated in Figure 5.2, the differences in construction cost per certification level were investigated, and three findings were observed:

- There are similarities among the low-cost projects.
- There are similarities among the Silver level projects.
- The high construction cost of Certified level buildings may be related to the higher security and safety requirements, rather than the sustainability pursuit.

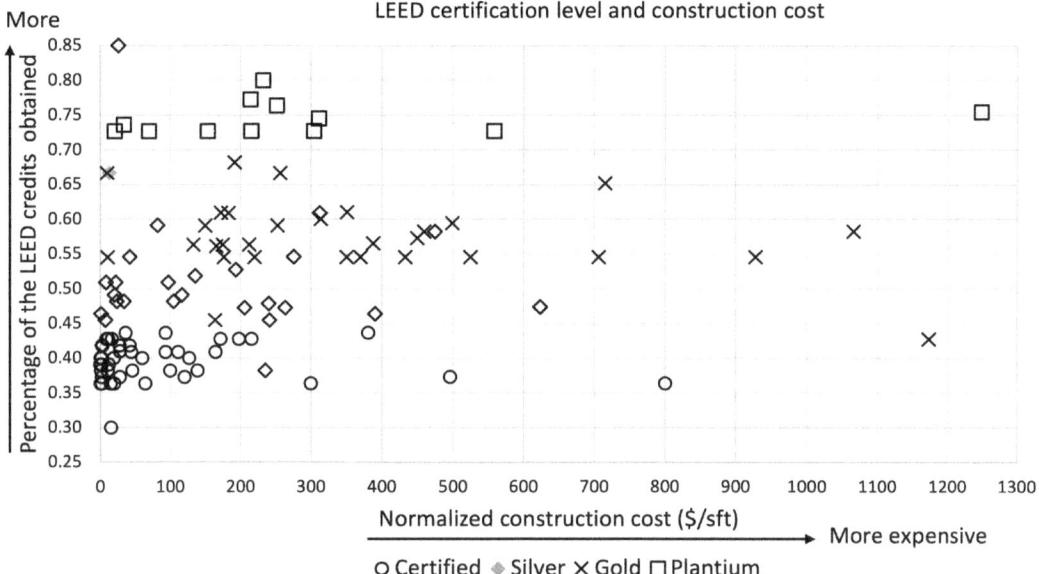

Figure 5.1
Construction cost and LEED certification level

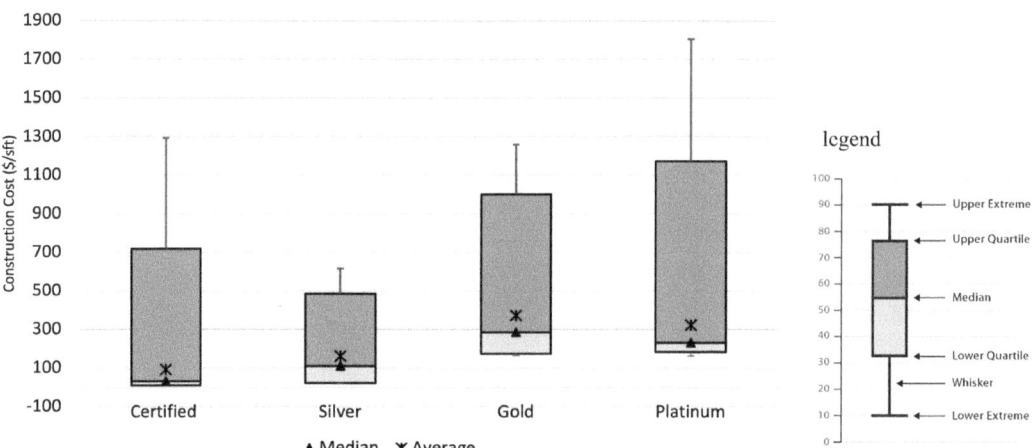

Figure 5.2
Construction cost per certification level

Next, a description of how to interpret Figure 5.2 is provided.

First, across all LEED certification levels, the dark gray boxes (Q3 tile) indicate a large cost variation among the projects, with the construction costs being above average. Conversely, the cost variance among projects with a lower construction cost than average is small, indicated by the shorter light grey boxes (Q1 tile). This large variation in expensive projects (above the median and average cost) and small variation in cheaper projects (projects have below-median and average costs) are further presented by the data in Table 5.3. The projects with a lower-than-average construction cost represent the majority (Certified = 68%, Silver = 57%, Gold = 63%, and Platinum = 82%). This observation from the data suggests that there are common practices or similarities among low-cost projects that lead to a similar normalized SBCC regardless of the size or location of the projects.

Second, the whiskers of the project costs at the Silver level are comparatively shorter than those of other certified levels (refer to Figure 5.2). Shorter whiskers indicate that the certified Silver projects have similar normalized SBCCs, which may be due to many reasons, one of which is that common practices exist in those Silver projects. Altogether, there are 30 Silver projects included in the data set, and 83% of the projects are full-scale renovations. Unlike the common perception of renovation projects having a higher uncertainty than new construction, the costs of those projects are much lower than their conventional counterpart. In Chapter 6, case studies are used to demonstrate how uncertainty and risk were well managed by the project team in those renovation projects, thus the construction overrun was well mitigated and controlled.

Third, the upper whiskers for the Certified and Platinum projects are much longer than those of Silver and Gold, which suggests there is a large cost variance within the top 25% (Q4 tile) of Certified and Platinum projects, most likely caused by few outliers with very high construction costs. For instance, the highest LEED Platinum project cost was around $17,276/m² ($1,605/ft²), which is a port of entry building. Compared to Silver level GSA office buildings, in addition to having a blast-resistant design, port of entry buildings have higher requirements for security and certain unique plan layout requirements. According to the US Land Port of Entry Design Guide, the types of spaces in land ports of entry include dry and wet laboratories (for food, drugs, cosmetics, biological products inspection), automated data processing centers, and other special spaces (Conway, 2021) that are not included in a typical commercial building. Those special spaces have a higher unit cost. According to Cumming Insight, a laboratory's cost in 2021 was about 22% higher than a normal commercial office (mid-rise) in the United States (Cumming Construction, 2021). Consequently, it is reasonable to speculate that the high cost is largely induced by the higher security and additional function requirements instead of by the sustainability pursuit.

Table 5.3 Percentage of projects regarding construction cost

		Certified	Silver	Gold	Platinum
Percentage of projects above the average construction cost					
		32%	43%	37%	18%
Percentage of projects below the average construction cost					
		68%	57%	63%	82%

Table 5.4 Regression analysis results: hard costs vs. soft costs

Variables		R-squared	ANOVA F signifi-cance	ANOVA F	p-value	Coeffi-cients
Hard costs		0.956	< 0.05	94.97	**< 0.05**	0.993
Soft costs	Design				**< 0.05**	0.906
	Management				0.122	0.420

Notes: Bold values represent the statistical significance. A result is considered statistically significant if the probability of it occurring by chance is less than a certain value, usually 0.05. This means that if the experiment were repeated many times, the result would only occur by chance 5% of the time or less.

5.1.3 Influential factors: soft costs vs. hard costs

Hypothesis 1b (H1b) states that the primary components contributing to the higher cost of sustainable buildings are derived from the hard costs. The hypothesis was rejected. As illustrated in Table 5.4, the regression model shows hard costs and design costs were both statistically significant ($p < 0.05$) and independent of each other. Within the soft costs, the design cost was significant ($p < 0.05$), while the influence of the cost of management and inspection was not statistically significant to the construction cost ($p > 0.05$). **Consequently, 95.6% of the construction cost variance can be explained by changes in hard costs and design costs, which have similar levels of influence**.

These findings are critical and demonstrate that although the design cost is a very small portion of the total construction cost, ranging from 2% to 19% in the studied projects, it has the same impact level on the total construction cost variance as that of hard costs. The explanation of these findings aligns with the effort curve described in Chapter 1, which states that the most important decisions that influence the construction cost are often made by the design team in the early stages and reflected in the design cost. To further understand the drivers and factors related to design cost, project team characteristics and project factors extracted from the SBCC database (refer to Chapter 3 for a description of the project characteristics and project team characteristics) are used to construct a structural equation model (SEM), which is explained in the following sections.

5.2 STRUCTURAL EQUATION MODEL

An SEM is a comprehensive statistical technique that examines the hypotheses of the relationships between observed and latent variables. Observed variables are variables that can be measured and recorded, while latent variables cannot be directly measured nor recorded. The latent variable, also referred to as a factor or construct (in research terms), is often hidden (Glen, 2020). For example, the quality of life is a latent variable, which can be inferred by a set of observed

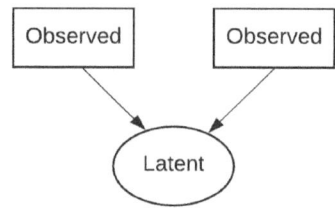

Figure 5.3
Graphic
representation of
variables

variables, including wealth, employment, physical and mental health, education, and social connections. Observed variables can also be subclassified with other variable types, with the most common being observed exogenous variables (equivalent to an independent variable) and observed endogenous variables (equivalent to a dependent variable). As illustrated in Figure 5.3, in an SEM diagram, an observed variable is represented by a rectangle and a latent variable is represented by an ellipse.

The development and adoption of SEM can be traced back to the 1900s, due to the growing needs of academics and practitioners in the social sciences who were looking for an effective way to understand the hidden factors and interactions of the factors behind human behavior (Tarka, 2018). Since then, SEM has been adopted by many disciplines and used as a multivariate analysis method to examine causal relations between observed and latent variables (Cho et al., 2009). For example, the use of SEM is now common in the natural sciences, especially in ecology and evolutionary biology (Mitchell, 1992; Sales et al., 2015).

In short, SEM is the merging of two analytic approaches: factor analysis and path analysis (Tarka, 2018). Charles Spearman, a British psychologist, is credited with developing the statistical technique known as factor analysis. He laid the foundation for SEM by constructing the first factor model, which later became an important measurement component of SEM (Spearman, 1904). Path analysis and systems of simultaneous equations were developed in genetics, econometrics, and later sociology. Seawall Wright, an American geneticist, known for his work on evolutionary theory, is credited with developing path analysis in the 1960s. The two approaches were merged in the early 1970s. Hauser and Goldberger (1971) worked to include the unobservable into path models, which later became defined as latent variables. Jöreskog (1970) developed a general model for fitting systems of linear equations and for including latent variables. Further, in the early 1970s, the first computer program was developed by Jöreskog, Gruavaeus, and van Thiilo, named ACOVS, which was virtually a precursor of a highly popular computer software currently in use, LISREL (officially in use in 1972) (Jöreskog & van Thiilo, 1972).

5.2.1 SEM in built environment research
SEM has been used as an effective and versatile research tool to study the built environment quality, satisfaction, and related occupant behavior, both indoors

and outdoors. Leung et al. (2020) used SEM to study the relationship between the indoor built environment and behaviors of residents with dementia in care and attention homes in Hong Kong. Five structural models were created based on the correlation identified through initial specification using regression analysis to study the indoor built environment's impact on emotion, mobility, sleeping disturbances, and loneliness. Kamaruzzaman et al. (2015) used the SEM approach to measure occupants' level of satisfaction and the perceived importance of the design of an office refurbishment. Twenty-two factors were included in the SEM, and the analysis results showed the appearance of the office conditions as the most significant influence on occupants' satisfaction. Da Silva et al. (2020) applied the confirmatory factor analysis and multiple linear regression analysis to evaluate apartment owners' perceptions regarding the built environment quality, the perceived quality of services, and their satisfaction. The researchers found the factors in management were strongly correlated with higher levels of apartment owners' satisfaction. Tekce et al. (2020) developed a hierarchical structural model to study office building occupants' satisfaction based on a literature review, analysis of occupant feedback records, and focus group meetings. They used SEM to validate this model through survey data collected from 300 office occupants. The proposed SEM included 27 factors across six constructs. The analysis results showed building design (e.g., the layout, furniture, exterior design, amount of space, and interior design) and facility management services as the most influential in occupant satisfaction. Fakhari et al. (2021) used SEM to confirm that 12 studied visual variables (e.g., glare, lighting level) all significantly contributed to the overall visual comfort in a classroom setting. The data were extracted from 192 survey responses from high school students in Tehran, Iran.

Overall, SEM's application in built environment-related studies has been increasing in the past decades, both on an individual building scale and large urban scale.

5.2.2 SEM in construction research

The introduction of SEM to construction research is relatively new, but its application has been increasing (Xiong et al., 2015). One of the earliest published works used SEM to study the mediation effects of relational bonding on construction project performance (Sarkar et al., 1998). The performance was measured by economic criteria, as well as relational satisfaction, and the relational bonding (latent variable) was measured by four observed variables (e.g., termination cost). Using SEM, they tested the hypothesis that project performance is directly impacted by relational bonding as well as indirectly through the mediating influence of shared decision-making and relational processes. In another study, Molenaar et al. (2000) used SEM to investigate the effects of factors on contract disputes between contractors and owners. In both cases, SEM was used as a conceptual framework, as well as an analysis technique, to deepen the understanding of or explore the unknown aspects in construction research.

Another way SEM has been used in construction research is to study cost-related topics. For example, Abulhakim and Adeleke (2019) used SEM to examine the significant impact of materials, equipment, and workplace conditions on accident occurrence in construction projects. Wong and Cheung (2005) used SEM to assess the relation between trust and partnering success in construction projects, which is measured by the cost and schedule control. Aibinu and Al-Lawati (2010) developed a theoretical structural model representing the impact of six latent variables on the willingness of construction organizations to participate in e-bidding. Doloi et al. (2011) applied SEM for assessing the impact of contractors' performance on project success using survey data collected across medium-sized construction projects in Australia; they found that the technical planning and controlling expertise of the contractor was key to achieving a successful project, and the success was primarily measured by the cost and schedule control.

However, for the application of SEM specifically to the sustainable building construction cost, there are very few existing studies. In addition, most SEM studies use qualitative data collected through surveys, with few studies using empirical data collected from actual built buildings. One previous study from Cho et al. (2009) is relevant to the study the author conducted in this book. The team applied SEM to examine the effect of project characteristics on construction project performance; 17 project characteristics were grouped into five project performance constructs. The important methodological contribution is that this study established an SEM based on quantitative data from actual case studies as opposed to previous SEM studies that mainly used qualitative survey data.

5.3 SEM RESEARCH METHODOLOGY

5.3.1 Concept of SEM and use in the proposed work

The objective of this SEM analysis is to identify the causal relationship between the SBCC and the project characteristics and project team characteristics by executing a quantitative data analysis on built GSA LEED projects. The project characteristics and project team characteristics are latent variables measured by multiple observed variables.

The proposed SEM is comprised of two components: a measurement model and a structural model (refer to Figure 5.4). As shown in Figure 5.4, the measurement model represents the concept or theory proposed in the study. It includes exogenous observed variables (illustrated by B1–B13), and it specifies how the characteristics are measured by the indicators. Here, the characteristics are also called constructs, which are often latent variables. For example, the project team characteristic is a latent variable that is reflected in six observed variables (B1–B6). Also illustrated in Figure 5.4 and Table 5.5, the structural model tests the causal relationship between exogenous and endogenous variables (path between Y1, Y1 to project team characteristics and project characteristics). Figure 5.4 also lists the test hypothesis and associated variables (H2a–H2h). The final description and testing results are summarized in Figure 5.5 and Table 5.10.

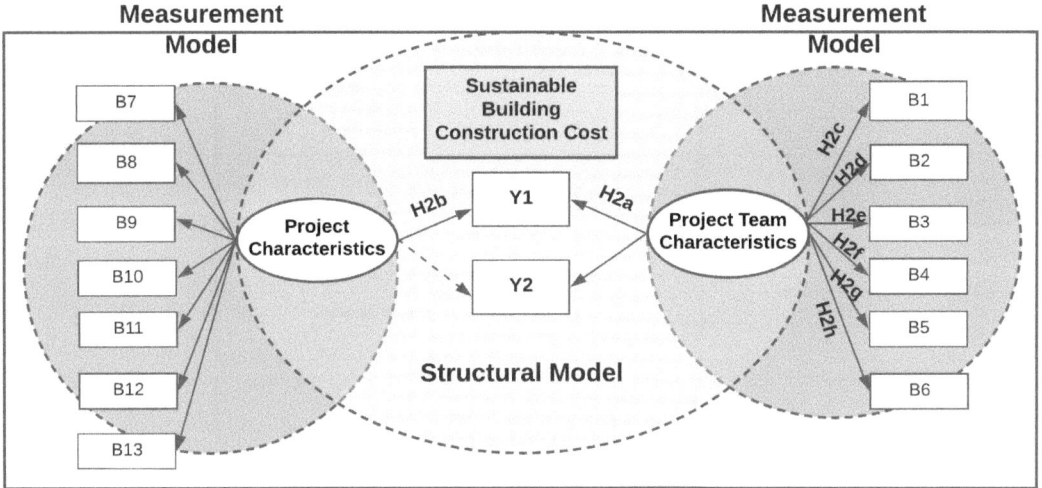

Figure 5.4
Specified SEM and tested hypothesis (H2a–H2h)

The overall structure of SEM can be expressed in the following fundamental equations (Bollen, 1989):

$$\eta = B\eta + \Gamma\xi + \zeta \qquad \text{(Equation 5.1)}$$

Equation 1 represents a structural model, where η signifies the endogenous variables, η and ξ represent the exogenous variables. B and Γ are coefficient matrices, and ζ represents the latent errors in the equations.

$$y = \Lambda\eta + \varepsilon \qquad \text{(Equation 5.2)}$$

$$\chi = \Lambda\xi + \delta \qquad \text{(Equation 5.3)}$$

Equations 5.2 and 5.3 represent measurement models for x and y, where Λ is the coefficient between γ and η and between χ and ξ. ε and δ are the errors in the equations.

Several steps are involved for developing, identifying, and testing the model. The purpose of the test is to confirm the ability of the proposed SEM conceptual model fit with the collected data. The following steps and testing are taken to identify the optimized model fit with the collected data.

5.3.2 Model development: measure variables

Table 5.5 shows all variables were evaluated based on nominal and/or ordinal scales, from one to five, in the third columns. The measurements of B1–B6 were explained in Chapter 4; B7–B9 and B11 are straightforward as described in Table 5.5. B10, B12, and B13 are explained as follows.

Table 5.5 Variables affecting the sustainable building construction cost

No.	Variables	Description
Project team characteristics		
B1	Procurement	1 = best value, 2 = low bid, 3 = sole source selection, 4 = qualifications-based
B2	Skills	1 = no skills, 2 = few skills, 3 = moderate skills, 4 = specialist, 5 = leading experts
B3	Experience	1 = no experience, 2 = little experience, 3 = average, 4 = experienced, 5 = exemplary
B4	Communication	1 = very poor, 2 = below average, 3 = average, 4 = above average, 5 = excellent
B5	Collaboration	1 = very poor, 2 = below average, 3 = average, 4 = above average, 5 = excellent
B6	Motivation to innovate	1 = very poor, 2 = below average, 3 = average, 4 = above average, 5 = excellent
Project characteristics		
B7	Level of sustainability	1 = LEED Certified, 2 = LEED Silver, 3 = LEED Gold, 4 = LEED Platinum, 5 = zero energy certified
B8	Building type	1= office, 2 = courthouse, 3 = port of entry, 4 = post office, 5 = inspection
B9	Construction type	1 = new construction, 2 = full renovation, 3 = limited renovation, 4 = restoration, 5 = others
B10	Technical complexity	1 = 0.5< ~ < 0.9, 2 = 0.91< ~ < 1.5, 3 = 1.51< ~ < 2, 4 = 2.1< ~ < 2.84, 5 = > 2.85
B11	Project location	1 = West, 2 = South Atlantic, 3 = Northeast 4 = South, 5 = Midwest
B12	Project scale	1 = 1< ~ < 50,000 ft^2, 2 = 50,001 < ~ < 200,000, 3 = 200,001 < ~ < 500,000, 4 = 500,001< ~ < 1,000,000, 5 = > 1,000,000
B13	Material availability	1 = 2 points, 2 = 3–4 point, 3 = 5–6 points, 4 = 7–8 points, 5 = > 8 points

Technical complexity (B10) was measured using two measurements. The first is according to points received in the LEED Energy and Atmosphere category since this category is considered a hallmark of a high performance building and a feature expected of LEED buildings by the broader real estate market (Winters et al., 2014). There are 35 points available, including three required points. For example, the Wayne Aspinall Federal Building and US Courthouse is certified LEED Platinum. It received 33 points, so the normalized score is

33/35 = 0.94. The second measure is the sophistication of the building system of studied projects. For instance, some projects have a conventional façade system, and others have a double skin façade or automated façade. There is a maximum of five points, with one being the most conventional and five being the most sophisticated. Using the Wayne Aspinall Federal Building and US Courthouse as an example, the average of the two scores is 0.94 + 5 = 2.97. It is more than 2.85 (refer to Table 5.5), thus the final score is five.

The project scale (B12) was evaluated based on nominal and ordinal scales. If the project total gross area was less than 50,000 ft², then it was given a score of one on the ordinal scale; 50,001–200,000 ft² received two points; 200,001–500,000 ft² received three points; 500,001–1,000,000 ft² received four points, and over 1,000,000 ft² received five points. For team skills (B2), it was given a rating based on the ordinal scale, consisting of one (no skills) to five (leading experts) points.

Material availability (B13) was used as a proxy for supply chain maturity and was measured by the points the project received under the LEED Material category. There are ten points available, including two required points. A project that obtained the minimal required two points was given a score of 1; 3–4 points received a score of 2; 5–6 points received a score of 3; 7–8 points received a score of 4; and >8 points received a score of 5.

5.3.3 SEM model estimation and evaluation

The last step was the model estimation and evaluation. Model estimation is the process of computing the best possible set of estimated variables to represent the given data. The default and commonly used estimator in SEM is maximum likelihood estimation (MLE), and the model assumes multivariate normality and that the variables are standardized. STATA was used to conduct model estimation and hypothesis testing. After the model was estimated, three primary criteria were used to evaluate the model and estimation results: (a) the fitness of the model (GOF), (b) whether there were negative variance problems (Haywood cases), and (c) whether the path and each parameter estimation result could be interpreted. The explanation for each criterion can be found in Section 5.4.3.

5.4 SEM ANALYSIS FINDINGS AND DISCUSSION

5.4.1 Correlation analysis

A Pearson correlation analysis was first conducted to estimate the interrelationships among the variables (refer to Table 5.6). The statistical significance of correlation was determined at $p < 0.05$; it appears as an asterisk (*) next to the correlation value. A positive value denotes a positive linear correlation, and a negative value denotes a negative linear correlation. The closer the value is to −1 or +1, the stronger the influence and correlation. The goal in this analysis was to identify the relation between project characteristic factors and project team characteristics. To interpret the degree of correlation, a high degree of correlation is between 0.7 and 1 (1 is a perfect correlation), a moderate correlation is between 0.3 and 0.69, and a low degree (weak) of correlation is between 0 and 0.29 (Ratner, 2009). In the analysis, the correlation between B1 and B2 was 0.5831; with an asterisk, it means there is a moderate positive correlation between B1 and B2. Table 5.6 demonstrates significant relationships among several variables, with three observations of importance:

A. LOS (B7) has a moderately positive relation with team skills (B2), experience (B3), and innovation (B6).
B. Construction type (B9) has a moderately negative relation with the LOS (B7) and building type (B8).
C. Team skills (B2) has a strong positive relation with team experience (B3) and a moderately positive relation with communication (B4), collaboration (B5), and innovation (B6); hence B2 could be viewed as a key variable to deciphering project characteristics.

5.4.2 Exploratory factor analysis and scale reliability test

After gaining an overall understanding of the correlation among variables, exploratory factor analysis (EFA) was used to discover the underlying factors (constructs) of the variables measured in this study (refer to Table 5.7). The sample-to-item ratios were higher than 10:1, meeting the recommended minimum requirement (O'Rourke et al., 2013). By using the principal factor analysis with varimax rotation, three factors were identified with Eigenvalues greater than 1, which should be retained. The three factors were *project team characteristics*, *project characteristic one*, and *project characteristic two*. The KMO values for project team characteristics, project scale, and project type are 0.660, 0.552, and 0.612, respectively, which are all greater than the recommended minimum value of 0.5 (Kaiser, 1958). Among the initial 13 variables, the factor loadings of 11 variables were higher than 0.5; variables B11 and B13, which had a loading factor less than 0.5, were removed from further analysis (Field, 2005).

After removing the insignificant variables (B11 and B13), a reliability test was conducted to identify the internal consistency of the variables within each factor using Cronbach's alpha value (refer to the sixth column in Table 5.7). The acceptable threshold of α is 0.7. The *project team characteristics* and *project characteristic one*, have an α value of 0.8047 and 0.7476, respectively, while the value of *project characteristic two* is below the threshold. However, when the author calculated the remaining variables under *project characteristics one* and *two* together, considering them as project characteristics, the α value was higher than the requirement. Therefore, B7, B8, B9, B10, and B12 were grouped together to form the new *project characteristic* category in the next step for hypothesis testing.

Table 5.6 Pearson correlation matrices

			B1	B2	B3	B4	B5	B6	B7	B8	B9	B10	B11	B12	B13
Project team character	Procurement	B1	1												
	Skill	B2	0.5831*	1											
	Experience	B3	0.5330*	09141*	1										
	Communication	B4	0.3295*	0.4971*	0.5523*	1									
	Collaboration	B5	0.035	0.4365*	0.3424*	0.2715	1								
	Innovation	B6	0.182	0.4507*	0.4120*	0.3555*	0.247	1							
Project characters	LOS	B7	0.288	0.3201*	0.3621*	0.2927	-0.083	0.3365*	1						
	Building types	B8	-0.069	-0.168	-0.108	0.0227	-0.216	-0.1781	0.0637	1					
	Construction types	B9	0.033	-0.023	-0.107	-0.1185	-0.006	-0.2374	-0.3729	-0.381	1				
	Technical complexity	B10	0.185	0.203	0.115	0.007	-0.061	0.2445	0.0978	-0.098	0.081	1			
	Location	B11	0.028	0.08	0.062	-0.0305	-0.111	-0.1482	-0.15	0.008	0.258	-0.04	1		
	Project scale	B12	0.3783*	0.237	0.152	0.0162	0.277	-0.0239	-0.063	-0.226	0.218	0.3331*	-0.082	1	
	Supply chain	B13	0.022	0.0084	0.018	0.0494	0.028	0.1155	0.5779*	-0.3242	-0.218	-0.05	-0.116	0.037	1

Table 5.7 Exploratory factor analysis of the variables

Principal factor	Nature	Code	Variables	Factor loading	α
Overall KMO = 0.608					
Project team characteristics					**0.8047**
(KMO = 0.660)	+	B1	Procurement	0.650	
	+	B2	Skills	0.900	
	+	B3	Experience	0.876	
	+	B4	Communication	0.653	
	+	B5	Collaboration	0.636	
	+	B6	Innovation	0.561	
Project characteristic one					**0.7476**
(KMO = 0.552)	–	B7	LOS	0.612	
	+	B12	Project size	0.553	
	+	B9	Construction type	0.763	
Project characteristic two					
(KMO = 0.612)					0.517
	+	B13	Material	0.232	
	+	B8	Building type	0.603	
	+	B10	Technical complexity	0.735	
	+	B11	Location	0.435	
Note: α = Cronbach's alpha value, KMO = Kaiser-Meyer-Olkin					
'+' indicates positive variables, '–' indicates negative variables					

Notes: Bold values represent the statistical significance. A result is considered statistically significant if the probability of it occurring by chance is less than a certain value, usually 0.05. This means that if the experiment were repeated many times, the result would only occur by chance 5% of the time or less.

5.4.3 SEM fitness (GOF) and negative variance (Haywood case)

Five common model fit indices were used to test the fitness of the proposed model. The indices are the chi-square ratio, root mean square error of approximation (RMSEA), p of close fit (pclose), comparative fit index (CFI), and Tucker-Lewis index (TLI). The models with at least four fit indices meeting the requirements were accepted in this study (Kline, 2015). Table 5.8 lists the test value and acceptable threshold for each statistic used.

To accept models with a chi-square test, they must have a p-value > 0.05 since the significance of a chi-square test is to indicate a poor fit of the model.

Table 5.8 Model test and indices value

Fit statistic	Test value	Acceptable threshold
Chi-square	0.153	$p > 0.05$
RMSEA	0.05	< 0.1
Pclose	0.223	$p > 0.05$
CFI	0.956	> 0.9
TLI	0.923	> 0.9

In the initial model fitness test, the p-value was 0.153, which is higher than 0.05, thus the model seems acceptable. The lowest possible RMSEA is 0. Values < 0.5 are considered indicative of a close model fit, and values up to 0.08 are acceptable (MacCallum, et al., 1996). The pclose tests whether the model departs significantly from one that is a close fit to the data (i.e., RMSEA $<$ or $= 0.05$). In this test, RMSEA was less than 0.05, and the pclose was not significant ($p > 0.05$). Both findings indicate a close model fit to the data. The CFI and TLI are both incremental fit indices. Values > 0.95 for these indices indicate a very good model fit (Schumacker & Lomax, 2004). Values of 0.90 or above are considered an acceptable model fit (Pituch & Stevens, 2016). In this test, CFI (0.956) and TLI (0.923) were larger than 0.90; both indicate an acceptable model fit. In addition to validating the acceptance of the proposed model, the coefficient value was also examined to ensure there was no negative variance (Haywood case) estimated, hence there was no need to respecify the model.

5.4.4 SEM model specification

As shown in Figure 5.5, the final SEM was developed using the results from the Pearson correlation and EFA to further investigate the relationship between the project team characteristics and project characteristics and the SBCC. The final SEM model consists of a structural component and measurement components. The structural component includes two observed variables (Y1, Y2) and two latent variables (project team characteristics and project characteristics). The two-measurement component includes the observed variables (B1–B12) and two latent variables, which is the overlap between the structural model and measurement model. Together, the group of variables explains the causal relationship between various project characteristics and the construction cost, and the relationship between project team characteristics and the construction cost. Most project characteristics included in this study are predetermined (exogenous variables), especially for the retrofit and renovation projects. For example, the project site (B12) and building type (B8) are inherited from the existing condition, which will not change. Technical complexity (B10) and construction type (B9) are influenced mainly by the existing condition and associated with the renovation strategies and sustainability goal. The LOS (B7) is related to the existing building condition but not necessarily determined by it. Lastly, the project characteristics (B1–B6) are independent of the project team characteristics and are exogenous variables.

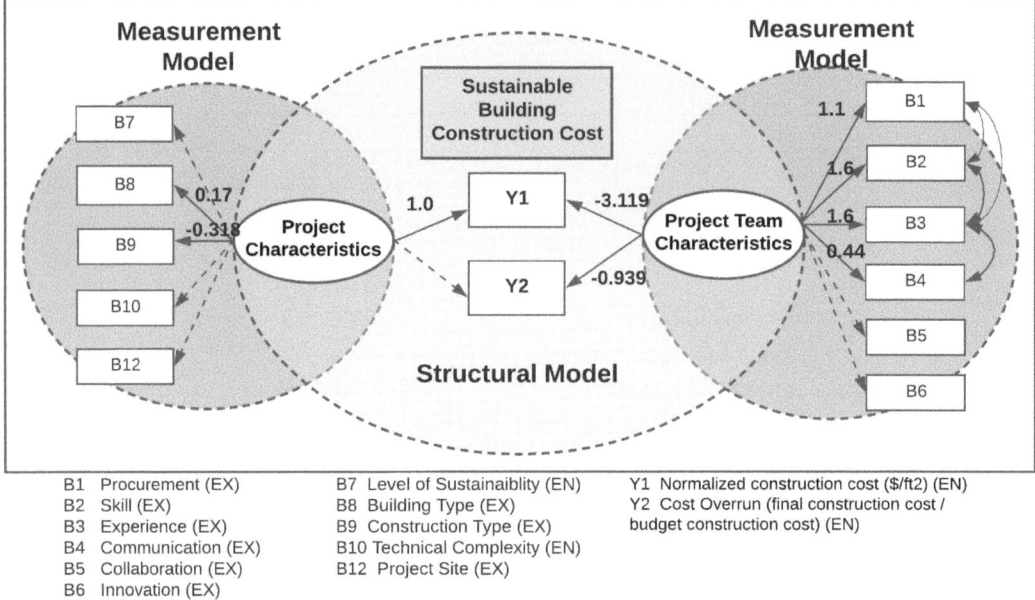

B1 Procurement (EX)
B2 Skill (EX)
B3 Experience (EX)
B4 Communication (EX)
B5 Collaboration (EX)
B6 Innovation (EX)

B7 Level of Sustainaiblity (EN)
B8 Building Type (EX)
B9 Construction Type (EX)
B10 Technical Complexity (EN)
B12 Project Site (EX)

Y1 Normalized construction cost ($/ft2) (EN)
Y2 Cost Overrun (final construction cost /
budget construction cost) (EN)

Figure 5.5
Final SEM model

The final model consists of 11 observed variables B1–B12, excluding B11, Y1, and Y2. The SBCC was represented by two observed endogenous variables, Y1 and Y2. Y1 is the normalized construction cost ($/m²), and Y2 is the cost overrun (final construction cost/budget construction cost); both are endogenous variables.

Moreover, each arrow in Figure 5.5 represents a coefficient showing the level of influence in each causal relationship. In general, **a path coefficient** is interpreted exactly as a regression coefficient, which is widely used to explain the level of influence. In addition, Figure 5.5 uses **dashed lines** to illustrate the causal relation, with a low level of confidence and no coefficient value. The **doubled-headed curved arrow** connecting observed variables (e.g., B1 and B2) indicates covariance existing among different variables. Covariance is not a causal relation; a positive covariance means the two variables move together, while a negative covariance means they move inversely. For example, being overweight raises the risk of developing high blood pressure. Consequently, a high blood pressure risk is negatively related to weight. They move in opposite directions but do not always have a direct causal relation.

5.4.5 Relationship between the project team characteristics and the sustainable building construction cost

Eight hypotheses were tested with SEM (refer to Table 5.9): H2a–H2h. Six were accepted, while H2g and H2h were rejected.

As shown in Figure 5.5 and Table 5.10, in the structural component (H2a), the project team characteristics negatively affect the unit cost (Y1, standardized

Table 5.9 Hypothesis testing results

H #	Description	Results
H1a	No correlation between construction cost and level of sustainability.	Inconclusive
H1b	Primary components contributing to the higher cost of SBs are derived from hard costs.	Rejected
H2a	Project team characteristics have a negative effect on final SBCC.	Accepted
H2b	Project characteristics have a significant effect on final SBCC.	Accepted
H2c	Procurement method has a positive relation to project team characteristics.	Accepted
H2d	Skill level of design project team has a positive relation to project team characteristics.	Accepted
H2e	Experience level of design project team has a positive relation to project team characteristics.	Accepted
H2f	Communication within design project team has a positive relation to project team characteristics.	Accepted
H2g	Collaboration within design project team has a positive relation to project team characteristics.	Rejected
H2h	Motivation to innovate of design project team has a positive relation to project team characteristics.	Rejected

coefficient = -3.119, $p < 0.005$) and also negatively affect cost overrun (Y2, standardized coefficient = -0.939, $p < 0.005$). In the measurement component, (H2c–H2h), the project team characteristics are influenced positively by **procurement type** (B1, standardized coefficient = 1.1, $p < 0.005$), positively by **team skills** (B2, standardized coefficient = 1.56, $p < 0.005$), positively by **team experience** (B3, standardized coefficient = 1.63, $p < 0.005$), and positively by **team communication** (B4, standardized coefficient = 0.44, $p < 0.005$). However, team collaboration (B5, standardized coefficient = -0.939, $p > 0.005$) and team innovation (B6, standardized coefficient = 0.45, $p > 0.005$) were not shown to have a statistically significant association with the cost.

A possible interpretation of the results is described as follows: The *higher* the level of team skills (B2) and team experience (B3), the *lower* the risk of cost overrun, hence the *lower* the final SBCC, which makes sustainable building more comparable to conventional building in terms of unit cost (even lower than that of conventional building). An examination of the project team makeup of the studied projects found that those design-built teams who worked on projects with unit costs that were comparable to conventional building and without cost overruns were in the top five of ENR rankings (Top 100 Green Buildings Design Firms, Top 100 Green Building Contractors) as leading experts in sustainable building design and construction firms. For example, ARUP is listed as third in the 2020 Top 100 Green Buildings Design Firms (Engineering News-Record [ENR],

Table 5.10 Estimation of relationship between the construction cost and project team characteristics

Standar-dized	Structural component		Measurement component						
Exogenous variables	Unit cost (Y1)	Cost overun (Y2)	Project team characteristics						
Observed variables			B1	B2	B3	B4	B5	B6
Coef	−3.119	−0.939	1.1	1.60	1.61	0.44	0.44	0.44
Std. Err	3.365	0.139	1.3	0.03	0.31	0.35	0.64	0.49
p	**0.0035**	**0.004**	**0.000**	**0.000**	**0.000**	**0.000**	0.011	0.008

Notes: Bold values represent the statistical significance. A result is considered statistically significant if the probability of it occurring by chance is less than a certain value, usually 0.05. This means that if the experiment were repeated many times, the result would only occur by chance 5% of the time or less.

2020), and Hensel Phelps is listed as fifth in the 2020 Top 100 Green Building Contractors (ENR, 2020). Both firms have more than 10 projects together, with most being LEED Platinum.

The procurement type (B1) is closely related to the qualification of the firms, which is often measured by skills and experience, explaining why procurement is also positively related to the construction cost and closely correlated to team skills and experience.

Compared to team skills and experience, team communication (B4) has less influence on the construction cost. As described and explained in Chapter 4, communication is measured both quantitatively by the number of communication channels and qualitatively by the communication process, effectiveness, and style. This finding differs from the widely accepted concept of more communication being better. On the contrary, excessive communication through many communication channels can cause misinterpretation of information. Effective and accurate communication is valued over the frequency and volume of the communication. More skilled and experienced project teams may communicate less when their communication and coordination is effective. These results confirmed several previous studies' findings. For example, Cho et al. (2009) found that a high level of communication among project team members was actually associated with a cost overrun and schedule overrun.

Overall, the effects of project team characteristics can be interpreted as follows: the **higher the level of skill** sets of a project team, **the more experience the project team** has with sustainable building, and the more **effective communication** the team has, the **lower the construction cost** caused by the high LOS the project achieves. These findings are in line with the impact of project team characteristics on conventional construction costs.

5.4.6 Relationship between project characteristics and the sustainable building construction cost

As shown in Figure 5.5 and Table 5.11, in the structural components, (H2b), the project characteristics positively affected the unit cost (Y1, standardized coefficient = 1.00, $p < 0.001$), while project characteristics do not have a statistically significant association with cost overrun (Y2, standardized coefficient = −0.016, $p > 0.005$). In the measurement component, the project characteristics are influenced positively by **building type** (B8, standardized coefficient = 0.017, $p < 0.005$) and negatively by **construction type** (B9, standardized coefficient = −0.318, $p < 0.005$). Level of sustainability (B7, standardized coefficient = 0.018, $p > 0.005$), technical complexity (B10, standardized coefficient = −0.01, $p > 0.005$), and project location (B12, standardized coefficient = −0.015, $p > 0.005$) do not have a significant influence on the construction cost.

The **insignificant causal relationship** between the **LOS (B7)** and the **construction cost** is not contrary to the widely accepted observation of a correlation between sustainability and the construction cost. The correlation demonstrates that the two items typically occur together in a consistent pattern, but the two items do not necessarily have a causal relation. For instance, there is a correlation between ice cream sales and sunglasses being sold. They do not cause the sales of each other, but there is a consistent underlying factor that causes both events to occur. The LOS is commonly perceived to increase the construction cost through technical complexity, which leads to longer construction times, higher labor costs, and higher material costs. Since the findings show no causal relation between technical complexity and the construction cost variance in studied cases, the **causal effect of LOS and the construction cost cannot be established**. This conclusion aligns with the regression model analysis from Section 5.1.

Table 5.11 Estimation of relationship between the construction cost and project characteristics

Standarized	Structural component		Measurement component				
Endogenous variables	Unit cost (Y1)	Cost overun (Y2)	Project characteristics				
Observed variables			B7	B8	B9	B10	B12
Coef	1.00	−0.016	0.018	0.017	−0.318	−0.010	−0.015
Std. Err	1.00	0.0058	0.14	0.006	0.008	0.009	0.008
p	**0.00**	0.78	0.03	**0.004**	**0.00**	0.31	0.008

Notes: Bold values represent the statistical significance. A result is considered statistically significant if the probability of it occurring by chance is less than a certain value, usually 0.05. This means that if the experiment were repeated many times, the result would only occur by chance 5% of the time or less.

5.5 DIFFERENCES

The findings from the data analysis differ from previous studies in three perspectives. The **first** is whether **technical complexity** is a significant cost driver for the SBCC. As previous studies showed, the SBCC risk is mostly influenced by project scope, technical complexity, and quality of design as sustainable building rating tools must satisfy certification requirements. The empirical analysis conducted for this book indicates technical complexity is not a significant factor to the final construction cost. **Instead, a project team's skills and experience are the most influential and determining factors that drive the cost of sustainable building because the construction cost risk can be mediated and moderated by the project team's skills and experience**. The higher the project team's skills and the more experienced the project team is in sustainable building design, the better the team can mitigate potential risks and uncertainty during the construction phase, thus reducing the risk of a cost overrun and/or schedule overrun. An effective way to select a skilled and experienced team is by utilizing a procurement method that includes both qualification-based and best-value-based criteria.

The **second** is the importance of soft **design costs**, which are closely related to the project team skills and experience. Although the design costs associated with project team characteristics (paid to the project team) are a small portion of the total construction cost, they bear an important impact on the total project construction cost. With soft costs, as demonstrated in the hypothesis (H1b) testing results, the design fee is not only significantly related to the final total construction cost, but its impact (0.906) is very close to that of hard costs (0.993). This finding aligns with the **effort curve** (refer to Chapter 1): the most important decisions related to cost and cost risk are decided in the early design phase. Consequently, it can be speculated that greater investigation in the early design phase and more focus on finding a qualified design team can be equally effective as hiring qualified contractors in terms of controlling the construction cost.

The **third** is the insignificance of **innovation** to the construction cost. Unlike previous studies and common perceptions, team collaboration and innovation was not found to be influential to the construction cost. This finding from SEM seems counterintuitive to common perception and study results; therefore, in-depth case studies are conducted to further understand why innovation and collaboration are not influential, and in what mechanism the project design team's skills and experience directly contribute to the final SBCC.

5.6 CHAPTER SUMMARY

Chapter 5 presents quantitative data analysis results from regression models and SEM analysis. The data analysis includes 72 GSA LEED-certified buildings. Sustainable building has long been viewed as a type of technology advancement that ignores the fact that sustainable design is a universal design principle. The analysis reveals that these basic principles are in fact not more expensive.

In view of the lack of empirical evidence (actual sustainable building cost data) to support the claim of sustainable building being expensive, the analysis conducted in this book based on empirical data **differs** from common perception. To further help readers understand those different results, the author attempts to address those questions in Chapter 6 with case studies.

REFERENCES

Abulhakim, N., & Adeleke, A. Q. (2019). The factors contributing to accident occurrence on Malaysia building projects through partial least square structural equation modeling. *Social Science and Humanities Journal, 4*(1), 1096–1106.

Aibinu, A. A., & Al-Lawati, A. M. (2010). Using PLS-SEM technique to model construction organizations' willingness to participate in e-bidding. *Automation in Construction, 19*(6), 714–724. https://doi.org/10.1016/j.autcon.2010.02.016

Bollen, K. A. (1989). *Structural equations with latent variables* (Vol. 1–210). John Wiley & Sons.

Cho, K., Hong, T., & Hyun, C. (2009). Effect of project characteristics on project performance in construction projects based on structural equation model. *Expert Systems with Applications, 36*(7), 10461–10470. https://doi.org/10.1016/j.eswa.2009.01.032

Conway, B. (2021). *Land port of entry*. https://www.wbdg.org/building-types/land-port-entry

Cumming Construction. (2021). *An in-depth look at construction costs per square foot in the United States*. https://www.buildingcalhhs.com/media/55

Da Silva, M. B. C., Giacometti Valente, M., Petroli, A., Detoni, D., & Milan, G. S. (2020). Perceived quality of built environment, service, satisfaction and value in use, in the context of residential buildings. *Journal of Facilities Management, 18*(4), 451–468. https://doi.org/10.1108/JFM-05-2020-0032

Doloi, H., Iyer, K. C., & Sawhney, A. (2011). Structural equation model for assessing impacts of contractor's performance on project success. *International Journal of Project Management, 29*(6), 687–695. https://doi.org/10.1016/j.ijproman.2010.05.007

Engineering News-Record. (2020). *ENR 2020 top 100 green buildings design firms*. ENR. https://www.enr.com/toplists/2020-Top-100-Green-Buildings-Design-Firms-Preview

Fakhari, M., Vahabi, V., & Fayaz, R. (2021). A study on the factors simultaneously affecting visual comfort in classrooms: A structural equation modeling approach. *Energy and Buildings, 249*, 111232. https://doi.org/10.1016/j.enbuild.2021.111232

Field, A. (2005). *Discovering statistics using SPSS*. SAGE Publications.

Glen, S. (2020). *Observed variables: Definition from StatisticsHowTo.com: Elementary statistics for the rest of us!* https://www.statisticshowto.com/observed-variables/

Hauser, R. M., & Goldberger, A. S. (1971). The treatment of unobservable variables in path analysis. *Sociological Methodology, 3*, 81–117. https://doi.org/10.2307/270819

Jöreskog, K. G. (1970). A general method for estimating a linear structural equation system. *ETS Research Bulletin Series, 1970*(2), i–41. https://doi.org/10.1002/j.2333-8504.1970.tb00783.x

Jöreskog, K. G., & van Thiilo, M. (1972). Lisrel a general computer program for estimating a linear structural equation system involving multiple indicators of unmeasured variables. *ETS Research Bulletin Series, 1972*(2), i–71. https://doi.org/10.1002/j.2333-8504.1972.tb00827.x

Kaiser, H. F. (1958). The varimax criterion for analytic rotation in factor analysis. *Psychometrika, 23*(3), 187–200. https://doi.org/10.1007/BF02289233

Kamaruzzaman, S. N., Egbu, C. O., Zawawi, E. M. A., Karim, S. B. A., & Woon, C. J. (2015). Occupants' satisfaction toward building environmental quality: Structural equation modeling approach. *Environmental Monitoring and Assessment, 187*(5), 242. https://doi.org/10.1007/s10661-015-4447-0

Kline, R. B. (2015). *Principles and practice of structural equation modeling.* Guilford Publications.

Leung, M., Wang, C., & Wei, X. (2020). Structural model for the relationships between indoor built environment and behaviors of residents with dementia in care and attention homes. *Building and Environment, 169,* 106532. https://doi.org/10.1016/j.buildenv.2019.106532

MacCallum, R. C., Browne, M. W., & Sugawara, H. M. (1996). Power analysis and determination of sample size for covariance structure modeling. *Psychological Method, 1*(2), 130–149.

Mitchell, R. J. (1992). Testing evolutionary and ecological hypotheses using path analysis and structural equation modelling. *Functional Ecology, 6*(2), 123. https://doi.org/10.2307/2389745

Molenaar, K., Washington, S., & Diekmann, J. (2000). Structural equation model of construction contract dispute potential. *Journal of Construction Engineering and Management, 126*(4), 268–277. https://doi.org/10.1061/(ASCE)0733-9364(2000)126:4(268)

O'Rourke, N., Psych, R., & Hatcher, L. (2013). *A step-by-step approach to using SAS for factor analysis and structural equation modeling.* SAS Institute.

Pituch, K., & Stevens, J. (2016). Assumptions in MANOVA. Applied multivariate statistics for the social sciences. In *Analyses with SAS and IBM's SPSS* (pp. 219–261). Routledge/Taylor & Francis Group.

Ratner, B. (2009). The correlation coefficient: Its values range between +1/−1, or do they? *Journal of Targeting, Measurement and Analysis for Marketing, 17*(2), 139–142. https://doi.org/10.1057/jt.2009.5

Sales, M. V. S., Gama-Rodrigues, A. C., Comerford, N. B., Cropper, W. P., Gama-Rodrigues, E. F., & Oliveira, P. H. G. (2015). Respecification of structural equation models for the P cycle in tropical soils. *Nutrient Cycling in Agroecosystems, 102*(3), 347–358. https://doi.org/10.1007/s10705-015-9706-5

Sarkar, M., Aulakh, P. S., & Cavusgil, S. T. (1998). The strategic role of relational bonding in interorganizational collaborations. *Journal of International Management, 4*(2), 85–107. https://doi.org/10.1016/S1075-4253(98)00009-X

Schumacker, R. E., & Lomax, R. G. (2004). *A beginner's guide to structural equation modeling.* Psychology Press, Taylor & Francis.

Spearman, C. (1904). "General intelligence," objectively determined and measured. *The American Journal of Psychology, 15*(2), 201–292. https://doi.org/10.2307/1412107

Tarka, P. (2018). An overview of structural equation modeling: Its beginnings, historical development, usefulness and controversies in the social sciences. *Quality & Quantity, 52*(1), 313–354. https://doi.org/10.1007/s11135-017-0469-8

Tekce, I., Ergen, E., & Artan, D. (2020). Structural equation model of occupant satisfaction for evaluating the performance of office buildings. *Arabian Journal for Science and Engineering, 45*(10), 8759–8784. https://doi.org/10.1007/s13369-020-04804-z

Winters, D., Sigmon, J., & Burt, L. (2014). *The LEED plaque unpacked: What a decade of LEED project data reveals about the green building market.* ACEEE Summer Study on Energy Efficiency in Buildings. https://www.aceee.org/files/proceedings/2014/data/papers/6-637.pdf

Wong, P. S. P., & Cheung, S. O. (2005). Structural equation model of trust and partnering success. *Journal of Management in Engineering, 21*(2), 70–80. https://doi.org/10.1061/(ASCE)0742-597X(2005)21:2(70)

Xiong, B., Skitmore, M., & Xia, B. (2015). A critical review of structural equation modeling applications in construction research. *Automation in Construction, 49,* 59–70. https://doi.org/10.1016/j.autcon.2014.09.006

Part III

The path forward

6 Comparative case study

ABSTRACT

To further validate and interpret the regression and SEM analysis results (refer to Chapter 5), six case projects were studied. The research questions addressed in the comparative study include how team skills and experience impact the construction cost as well as how team communication, collaboration, and innovation are relevant to the construction cost. Furthermore, the comparative study offers insight into why technical complexity was not an important factor in the construction cost. Overall, this chapter aims to provide solid real-life evidence to debunk the misperceptions around sustainable building and its cost.

NOMENCLATURE

ARRA	American Recovery and Reinvestment Act of 2009
BIM	Building information modeling
BRF	Byron G. Rogers Federal Building and US Courthouse
CMc	Construction manager as constructor
DB	Design-build
DBB	Design-bid-build
EGWW	Edith Green-Wendell Wyatt Federal Building
EISA	Energy Independence and Security Act of 2007
EUI	Energy Use Intensity
FCS	Federal Center South Building 1202
GMP	Guaranteed maximum price
GSA	General Services Administration
GTML	George Thomas "Mickey" Leland Federal Building
HMM	Howard M. Metzenbaum US Courthouse
IPD	Integrated project delivery
NHPA	National Historic Preservation Act of 1966
SHPO	State Historic Preservation Office
WAF	Wayne N. Aspinall Federal Building and US Courthouse

6.1 COMPARATIVE ANALYSIS AND GENERAL GSA PRACTICE

Six projects were selected as exemplary of successful sustainability achievement: Wayne N. Aspinall Federal Building and US Courthouse (WAF), Edith Green-Wendell Wyatt Federal Building (EGWW), Howard M. Metzenbaum US Courthouse (HMM), Federal Center South Building 1202 (FCS), Byron G. Rogers Federal Building and US Courthouse (BRF), and George Thomas "Mickey" Leland Federal Building (GTML). Together, they offer insights into numerous project characteristics, design strategies, and processes that were used by the teams to achieve aspirational performance goals. The range of selected projects allow readers to understand the project team characteristics' influence in various contexts: a historic preservation with net-zero energy goals in Colorado, the renovation and expansion of an urban high-rise office building in Portland, the renovation project of the first of 40 GSA LEED-certified buildings in Cleveland, new construction on a previously contaminated site in Seattle, a high-rise renovation in Colorado, and a 22-story LEED Platinum project in Houston. When documenting a project's process and team characteristics, a comparative case study strategy is most effective; therefore, a consistent format is employed in this chapter to explain the findings. Table 6.1 includes a comparison of the primary project characteristics and information for the six studied cases.

The 30-year (1991–2021) average inflation rate (excluding recession years) was 4% for nonresidential building and 4.75% for residential building. During the fast-growing period of 2013–2015, the inflation rate exceeded 4% for non-residential building, even reaching 6%–8% in certain major metropolitan areas (Edzarenski, 2021). With a 6% annual inflation rate, the current costs of the six case projects—WAF, EGWW, HMM, FCS, BRF, and GTM – would be $6,033/$m^2$ ($543/ft^2$), $4,611/$m^2$ ($415/ft^2$), $5,033/$m^2$ ($453/ft^2$), $5,233/$m^2$ ($471/ft^2$), $3,722/$m^2$ ($335/ft^2$), and $6,200/$m^2$ ($558/ft^2$), respectively. These prices are still comparable to or lower than the current average construction cost of a non-LEED building on a similar scale with a similar scope of work. As mentioned in Chapter 4 (Section 4.3), in 2021, the average construction cost was $6,244/$m^2$ ($562/ft^2$) for a new mid-rise (non-LEED) office building and $7,333/$m^2$ ($660/ft^2$) for a new high-rise office building, while the average full renovation project construction cost was $5,000/$m^2$ ($450/ft^2$) and $6,922/$m^2$ ($623/ft^2$) for mid-rise and high-rise office buildings, respectively.

Prior to the American Recovery and Reinvestment Act of 2009 (ARRA) (DOE, 2009), the General Services Administration (GSA) had limited experience with design-build (DB) project delivery, but several GSA regions had been studying the potential use of alternative delivery types that were more integrated than the conventional design-bid-build (DBB) project delivery method (Cheng, 2015). ARRA was signed into law by President Obama in February 2009, lasting until 2014. The goal of ARRA was to jump-start the economy, hugely impacted by the economic recession, to create or save millions of jobs. Moreover, ARRA stipulated that funded projects were to meet the stringent energy and water

Table 6.1 Comparative table of studied projects

Project name	Overview	Cost	Level of sustainability	Location
Wayne N. Aspinall Federal Building and US Courthouse (WAF)	Three stories 3,740 m² (41,562 ft²) 2013 completion	$15,000,000 $4,011/m² ($361/ft²)	81/110 points LEED Platinum Net-zero energy	Grand Junction, Colorado
Edith Green-Wendell Wyatt Federal Building (EGWW)	18 stories 46,122 m² (512,474 ft²) 2013 completion	$141,500,000 $3,066/m² ($276/ft²)	84/110 points LEED Platinum	Portland, Oregon
Howard M. Metzenbaum US Courthouse (HMM)	Five stories 21,297m² (236,632ft²) 2005 completion	$44,600,000 $2,100/m² ($189/ft²)	47/110 points LEED Certified	Cleveland, Ohio
Federal Center South Building 1202 (FCS)	Three stories (new) 18,810m² (209,000ft²) 2012 completion	$73,314,341 $3 888/m² ($350/ft²)	61/110 points LEED Gold	Seattle, Washington
Byron G. Rogers Federal Building and US Courthouse (BRF)	18 stories 44,460m² (494,000ft²) 2013 completion	$194,295,701 $3 477/m² ($313/ft²)	66/110 points LEED Gold	Denver, Colorado
George Thomas "Mickey" Leland Federal Building (GTML)	22 stories 32,940 m² (366,000 ft²) 2015 completion	$91,843,000 (lower than original) $2,366/m² ($213/ft²)	85/110 points LEED Platinum	Houston, Texas

conservation requirements of the Energy Independence and Security Act of 2007 (EISA) at that time. EISA was signed by President George W. Bush in December 2007, with the aim of increasing buildings' efficiency (EPA, 2007).

The context of ARRA, with its short time frames and aspirational goals for economic stimulus and high-performance buildings, provided ideal conditions for testing the use of more collaborative delivery types, such as DB or integrated project delivery (IPD) (Cheng, 2015). Although the DB project delivery method as a collaborative method was new to GSA then, the DB method has been a well-established method in the private sector since 1993, when the Design-Build

Institute of America was founded. Since then, DB has steadily gained ground in residential and commercial buildings in the private sector (Park & Kwak, 2017). Because of the unfamiliarity with DB as an institution, initially, the GSA selection process and policies were not well aligned for this type of project delivery. However, these difficulties were relatively easy to overcome due to extensive experience with DB held by several GSA team members from the private sector, partner architects, and contractors (Cheng, 2015).

Compared to DB, IPD was a newer delivery type, and fewer people had experience with it; thus the lack of established procedures created flexibility for project teams. For instance, the project team for the EGWW project developed a customized delivery method based on IPD principles. This customized delivery method helped the project team not only meet cost budget constraints but also achieve an unprecedented energy performance goal. For all case projects, the adaptation of standard GSA practices to these newer delivery methods required additional time investment and support in the design phase, and the most important decisions were often made during the planning and design phases.

6.2 CASE 1: WAYNE N. ASPINALL FEDERAL BUILDING AND US COURTHOUSE

The Wayne N. Aspinall Federal Building and US Courthouse (WAF) was built in 1918 in Grand Junction, Colorado. It is a three-story office building with a gross area of 3,740 m^2 (refer to Figure 6.1a). It originally functioned as a post office and courthouse, and a large extension was added to the east side of the original building in 1938 for office space. The original design and construction did not include an HVAC system, which was added in the 1960s, along with an upgraded electrical system. Acoustic ceiling tiles were also added to improve the sound quality as well as conceal the HVAC duct work. The original lighting fixtures were replaced with ceiling-mounted fixtures and wall sconces (Cheng et al., 2014). Since then, there had been no major repairs, renovations, or upgrades. The building was listed on the National Register of Historic Places in 1980 (General Services Administration, 2014).

In 2010, GSA considered disposing of the building due to its extensive repair and renovation needs. At that time, GSA could not cover the potential costs from existing or expected federal funding appropriations. Most of the building systems and components had reached the end of their useful life and required major updates and repairs, and the poor ventilation and indoor air quality needed to be addressed as well. Disposal of the existing building was avoided when the building received funding from ARRA to undergo a major renovation. GSA created an unprecedented extensive renovation plan, combining a net-zero energy goal and historic renovation goal. The renovation was completed in 2013 and achieved LEED Platinum certification, with a score of 81/110. The total construction cost was about $15 million, and the unit cost was $4,011/m^2.

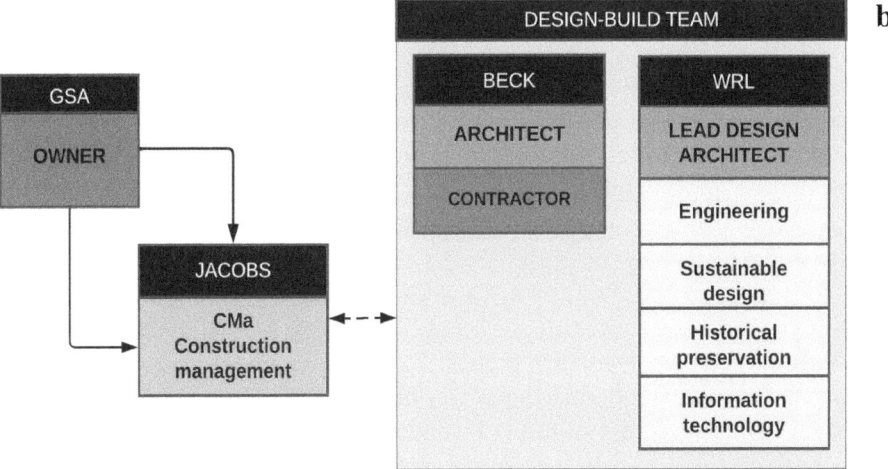

Figure 6.1
Wayne N. Aspinall Federal Building and US Courthouse: (a) after renovation with roof-mounted solar panels (photo by GSA) and (b) team organization

6.2.1 Technical complexity

6.2.1.1 Renovation while meeting existing tenants' requirements

An overarching challenge was undertaking the construction in an occupied build-ing, where construction crews could encounter unforeseen conditions and risks, causing construction delays that lead to a construction cost increase. There were situations when the project energy efficiency goal did not align with tenant pref-erences. For instance, the project team proposed consolidating copy rooms and server rooms to reduce the energy load of the building; however, the existing tenants preferred to keep them in separate spaces. While the project team was able to persuade tenants to agree on consolidated copy rooms, due to security concerns, the idea of consolidating server rooms was eventually abandoned.

6.2.1.2 Aggressive energy goal vs. historic preservation

According to the State Historic Preservation Office, the most controversial design element was the roof-mounted PV array (refer to Figure 6.1). Referring to Section 106 of the National Historic Preservation Act of 1966 (NHPA), the SHPO's initial review critiqued the appearance of the large PV panels in the view of buildings' front façade. It was said to "destroy historic... features and spatial relationships that characterize a property." The NHPA reviewers requested that the project team reduce or remove the PV canopy from the building.

Reducing the PV size or removing the PV canopy would have had a detrimental effect on meeting the net-zero energy goal, so the project team found an alternative solution: they increased energy efficiency through the geothermal system concurrently with increased insulation in the building envelope. A geothermal heat exchange system is about 45% more efficient than a typical HVAC system, and in this project, the vertical geothermal well was approximately 475 ft deep (Cheng et al., 2014). By combining a super-insulated building envelope and geothermal-integrated HVAC system, the project design team was able to reduce the heating and cooling demand and, consequently, the PV panel size while meeting the net-zero energy goal. A geothermal heat exchange system is a relatively advanced building system compared to a conventional HVAC system. The related technical complexity was mediated by the project design team, who had previous experience implementing similar systems in other projects. Therefore, the design, construction, and installation of this system was able to stay within the original budget.

6.2.2 Sustainability

GSA set high goals for this historic preservation project: (1) realization of net-zero energy (aligning with the government requirement for net zero and energy independence by 2030), (2) achievement of LEED Platinum certification, (3) improvement of indoor environmental quality and thermal comfort, and (4) reduction of water use by 40%. The modeled energy use intensity (EUI) was 14 kBtu/ft^2/yr, and the percentage of energy use reduction compared to the national median EUI for similar building types was around 78%. The designed lighting power density was 0.33 W/ft^2, which was much lower than the code-required 0.9 W/ft^2 at that time (AIA Top Ten, 2016).

6.2.3 Procurement and contract

GSA used a DB contract to bring the project team (design and construction) together from the outset of the project. WAF followed a two-step procurement process, combining qualification-based selection and best value selection. GSA first checked the qualifications of the potential teams and then issued a request for the proposal. The proposal phase required bidders (contractor and designer team) to address minimum performance criteria and encouraged bidders to pro-vide innovative design solutions. Since a bidder constitutes the design team and contractors, in their proposal, they can propose an integrative way to mitigate risk, consequently maintaining the pricing from design development, based on the contract documents, through construction.

In the WAF contract, the project team's design proposal was incorporated into the final contract. Even the PowerPoint presentation delivered by the team during the interview became a part of the construction documentation for the project. This highly integrated manner was thought to have helped the project mitigate potential risks encountered during the construction.

6.2.4 *Project team characteristics*

6.2.4.1 *Team skills and experience*

Selecting a team for the historic renovation project with a net-zero energy goal was the most important upfront decision. GSA first selected Jacobs Technology as the construction manager as advisor (CMa), then GSA selected a DB team based on a combination of past performance, technical capacity, and qualifica-tions related to skills and experience. As illustrated in Figure 6.1b, the DB team mainly comprised of Westlake Reed Leskosky (WRL) and the Beck Group.

Westlake Reed Leskosky (WRL, later acquired by DLR) provided MEP engi-neering, interior design, historic preservation, and LEED consulting services. WRL was founded in 1905 in Cleveland. Since 1997, WRL had redefined and narrowed its focus on building types, and historic preservation was one of its top focus areas. In 2016, WRL ranked seventh in *Architect* magazine's Top 50 US Architecture Firms (Cleveland.com, 2019). *Architect* has been the official publica-tion by the American Institute of Architects since before World War I. Its rankings are based on evaluations of architecture firms' business operations, sustain-ability practices, and design excellence. WRL's expertise in sustainable design was also supported by its strong commitment to research. In 2013, 60% of firm profits went back into research, which is rare for architecture and engineering firms (ARCHITECT, 2016). This investment helped to build the firm's technical strength, leading to successful projects. The Beck Group, a design and construc-tion firm based in Texas, was the general contractor and architect of record for the WAF project. It ranked 34th (in 2018) and 38th (in 2019) in ENR's Top 100 Green Building Contractors (ENR, 2019).

Both Beck and WRL were interdisciplinary firms with established cultures of working collaboratively between disciplines. Although these two firms did not have working relations prior to the project, their internal organizational cultures were compatible and required little alignment or adjustment, demonstrating that skills and experience were underlying factors for good communication and collaboration. Based on previous experience, both firms understood what were the most effective communication methods and how to create a col-laborative team. In addition, team members noted that they believed others would perform as promised and that each team member or firm would hold themselves responsible. This high level of trust was not based on previous working relations; rather, it was derived from the evaluation of members' and firms' skills and experience.

6.2.4.2 Team collaboration and communication

WAF had a normal team organization in the DB delivery, as illustrated in Figure 6.1. The main collaboration and communication occurred between WRL and Beck since most team members were drawn from those two firms. Prior to working together in the newly established project team, the two firms already had well-established methods of integration due to their multidisciplinary nature. Consequently, individual team members were already well versed in cross-disciplinary collaboration within their own firms.

In the early design phase, the team collaborated remotely; later in the construction phase, the core team members from both WRL and Beck were located on-site, in the basement of the building, with others traveling to the site regularly. The GSA project manager was also co-located in the same place. The team noted the increase in direct working relationships, and the ability to get to know each other on a personal basis strengthened communication and trust among the core team members. The team believe the co-location strategy helped support communication, collaboration, and management.

6.2.4.3 Innovation

The first innovation of this project was the method the project team used to manage risk and uncertainty. First, to manage risk, the project team contacted SHPO to mutually develop a strategy facilitating the historic review process. In this way, the project team was able to gain incremental progress and split the risk into pieces that could be managed. Second, to manage the uncertainty derived from the building's continued operations during the construction, the project team used physical mock-ups to effectively communicate the design intent to the project manager from GSA as well as to tenants. Mock-ups of the interior space renovation with intended layouts and furnishings allowed all parties to review and understand the design ideas. It should be noted that the above-mentioned innovation had no direct relation to the building's sustainability, nor was it created to achieve the sustainability goals. This innovation can be applied to any projects facing similar challenges, namely historic preservation, while renovating a building to meeting higher standards.

The second innovation was building information modeling (BIM) and its utilization. BIM was employed as a design, documentation, communication, management, and collaboration tool throughout the design and construction to manage project complexities. It was also employed to assist in design decision-making related to the net-zero energy performance goal. The project team developed a BIM execution plan at the beginning of the project to set the procedures of developing a BIM, share and update the documents, and outline each member's responsibility. BIM was also linked to energy models early in the project. Regardless of the perceived advantages of utilizing BIM in DB projects, the actual benefits remained somewhat unclear. As one team member noted:

> Within the project team, we still have to communicate the information scope out to the subcontractors. That information can be related in different ways, but the content still needs to be developed to a level that somebody outside the team can understand.

These undefined benefits contributed to why using BIM did not significantly impact the total construction cost of this project.

6.2.5 Construction cost

The WAF project had a fixed price budget which was set during the early stages. When the project team made a proposal to bid for the project, the scope and schedule were considered. The GSA project manager noted that the fixed price was a clear motivator for the team: "(We) did not want to go back and ask for more money. We had a team that was very good at identifying what we could do to make things cost-effective" (Cheng et al., 2014). The procurement method employed and the selection criteria integrating qualifications (design team skills and experience) and best values (design proposal) enabled GSA to select a design team who could deliver a project design meeting all energy performance goals within the allowable budget. This case study is a clear indication that a skilled and experienced design team not only provides practical design solutions but also knows what tools and methods can facilitate communication and collaboration, even during the construction phase.

6.3 CASE 2: EDITH GREEN-WENDELL WYATT FEDERAL BUILDING

The Edith Green-Wendell Wyatt Federal Building (EGWW) was built in 1974 in downtown Portland, Oregon. It is an 18-story office building with a total gross area of 512,474 ft^2 (refer to Figure 6.3a). Before the renovation, the building's mechanical system was worn out and outdated. The building received funding from ARRA to undergo a major renovation, with the scope including upgrading the building systems, modernizing work environments, and improving accessibility while meeting the energy and water conservation requirements of EISA. The renovation was completed within 39 months, in 2013, and the renovated building is a workplace for 16 federal agencies. This project achieved LEED Platinum certification, with a score of 84/110. The total construction cost was over $14 million, and the unit cost was $3,066/m^2 (278/ft^2). Unlike WAF, the EGWW was not occupied at the time of renovation. After the initial cost-benefit analysis, GSA decided to vacate the building for the retrofit since ARRA required funding to be spent by the end of September 2015 and vacating the occupants would accelerate the process.

6.3.1 Technical complexity

6.3.1.1 Complete building envelope retrofit

Replacement of the existing precast concrete cladding with a high-performance façade was one of primary challenges of the renovation (refer to Figure 6.2). The façade system had integrated aluminum reflecting and shading elements that were designed to respond to year-round variations in the sun's angle. On the south, east, and west sides, the new facade integrates a shading device to minimize the solar heat gain. On the south façade, the horizontal light shelf can reflect the light deeper into the office space, thus decreasing the demand for electrical lighting as well. Recladding the façade with a sophisticated layered system contributed somewhat to the sustainability goal, but its contribution is limited. Mainly, it is an added design feature since this project served as a flagship example to demonstrate design excellence.

6.3.1.2 HVAC system retrofit: radiant cooling system

Another major technical challenge was the installation of a hydronic radiant ceiling with a direct outside air system. It was estimated to save 10%–15% of the total building energy use when compared to a variable air volume mechanical system (refer to Figure 6.2c). Radiant cooling was a relatively new technology in the United States; therefore, to manage risk, the design team reallocated funds and designated a team member to work directly with the manufacturing facility.

6.3.2 Sustainability

EGWW is LEED Platinum certified. Before the renovation, the EUI of the existing building was 75.5 kBtu/ft^2, and the target EUI after renovation was 29 kBtu/ft^2, which is a 55% reduction from the national average, compared to buildings of a similar size and type (Cheng, 2015). The designed lighting power density was 0.60 W/ft^2. Together with enhanced daylight renovation, it can lead to a 50%–60% lighting energy reduction (AIA Top Ten, 2016). For water conservation, EGWW achieved 60% in water savings through strategies of incorporating both water-conserving plumbing fixtures, water reuse, and rainwater capture. One unique sustainable practice was a reduction in demolishing waste, which contributes to material conservation. A subcontractor in charge of demolition was brought onto the project team at the beginning of the design, which was unusual. The subcontractor worked with the project members to develop demolition drawings and specs as well as recycle and reuse strategies during the design phase. These practices served two purposes: (1) to help the demolition contractor understand the project's sustainability goals, and (2) to help the team to create a process with the demolition contractor to track the material conservation goal (Cheng, 2015).

6.3.3 Procurement and contract

Prior to receiving funding from ARRA, the GSA's Northwest/Arctic Region (region 10) hired SERA Architects for design services of the renovation project. The

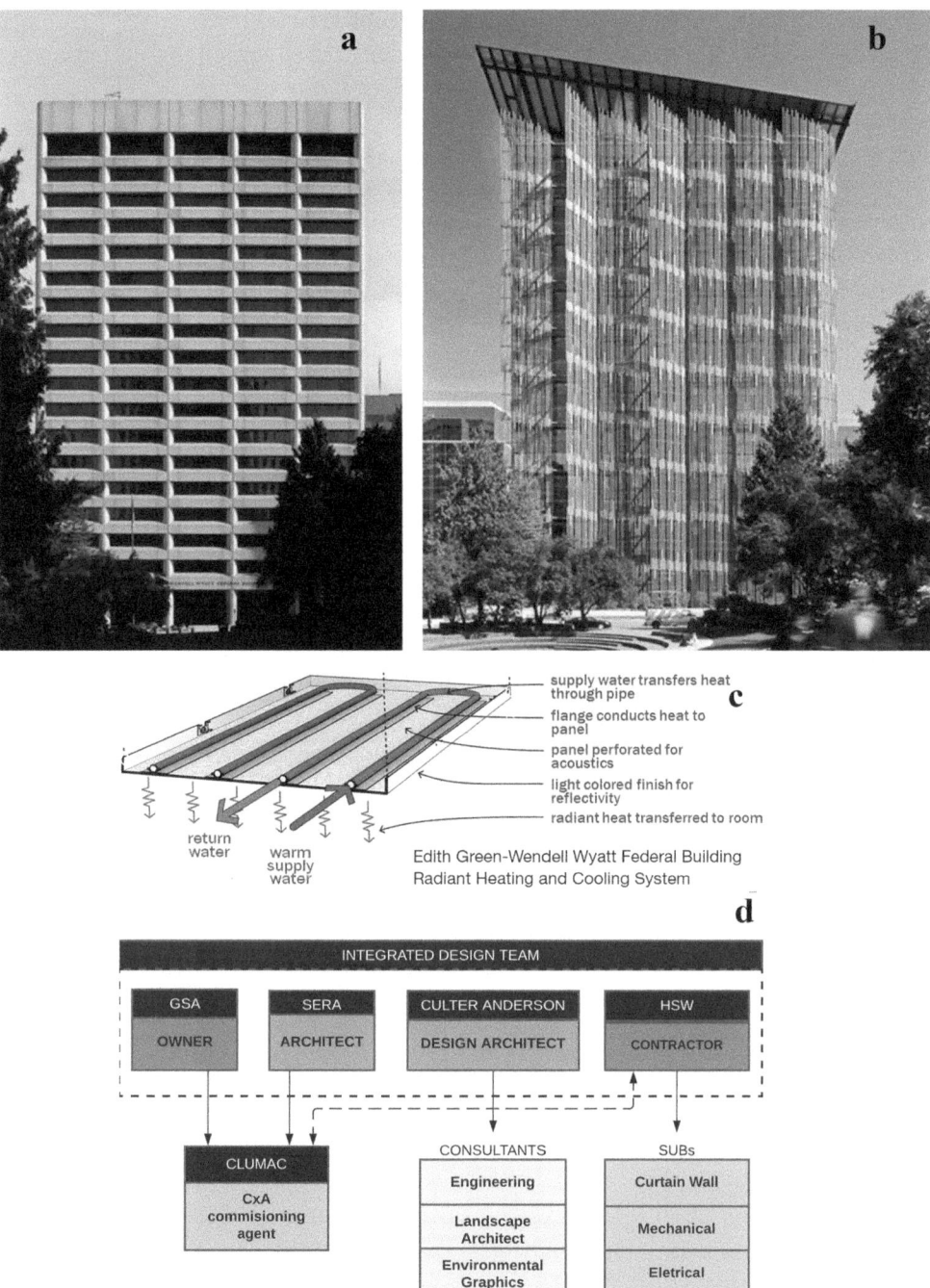

Figure 6.2
Edith Green-Wendell Wyatt Building: (a) before renovation, (b) after renovation, (c) radiant heating and cooling ceiling system (photo by GSA), and (d) EGWW project team organization (Cheng, 2015)

original contract was a traditional DBB delivery method. After obtaining approval for ARRA funding in April 2009, the EGWW project was required to be re-scoped to align with the high-performance green building goals, which added a vastly different set of goals and technical specifications. In addition, the ARRA funding required a commitment by September 2010 and had to be spent before September 2015. These defined funding requirements and time frame became the driving force to switch from a traditional DBB to more of an IPD method. After conducting market research, GSA determined that a general contractor/construction manager delivery method, which the GSA refers to as the construction manager as constructor (CMc), along with a guaranteed maximum price (GMP) contract type, was most appropriate for the project scope and constraints – specifically, the securing of project funding by September 2010 (Cheng, 2015). The final delivery method for EGWW was CMc using a GMP. CMc was not addressed in federal rules before 2018, and CMc/GMP is not currently addressed in federal rules; therefore, this type was authorized for use in EGWW as an exception (Cheng, 2015). In addition, unlike a normal CMc delivery method, the significant variation in EGWW was that the formation of design (architect and engineer) was early (prior to the request for proposal). After gaining ARRA funding, GSA decided to retain SERA Architects as the architect of the project and engaged the firm in the request for the proposal process. Among the reasons for retaining SERA Architects were the firm's expertise in high-performance green buildings and its experience.

Two unique aspects about the procurement and contracting process were the primary contractor team selection and subcontractor selection. EWGG invited all the general contractor candidates to a high-performance green building re-scoping workshop. The final selection of the contractor was based on technical capability, including key personnel's expertise and experience. The primary contractor was required to submit recommendations for five first-tier subcontractors, which was unusual for a GSA project. This process helped the project team and GSA management team ensure that the contractor team had the capacity to deliver the project. Moreover, bringing the subcontractor onto the project early could facilitate turning subcontractors into co-owners of the project, where they could be more involved in the value engineering process.

6.3.4 *Project team characteristics*
6.3.4.1 Team skills and experience
As mentioned above, SERA Architects was retained for its expertise in high-performance building design. In 2010, the prominent magazine for architectural professionals *Architect* ranked SERA Architects as one of the top three green design firms, followed by other large international firms, such as COOKFOX, HOK and FXCollaborative. At that time, 60%–79% of SERA's were LEED projects. Since 2008, SERA Architects has been developing the firm's in-house Sustainability Resource Group (*Architect* magazine, 2010), which comprises experts in different disciplines, including mechanical engineers, façade engineers,

daylight and lighting designers, and architects (*Architect* magazine, 2010). In this way, SERA Architects developed not only their expertise but also a collaborative mindset, approach, and culture for the project, which included effective communication and collaboration. GSA later hired Cutler Anderson Architects to work together with SERA as the design architect and architect on record to further develop construction documents (refer to Figure 6.2d).

The general contractor of the design team, Howard S. Wright (HSW), mainly had experience working within the Western United States. In 2008, the firm was ranked in the Top 100 Contractors by ENR, and it was acquired by Balfour Beatty in 2011. HSW had great experience in large and complicated projects, such as the Century 21 Exposition's Space Needle in Seattle, Washington, a 138-ft-tall observation tower that is currently a landmark building. The firm's experience and skills in dealing with complicated and non-precedent projects were the main reasons GSA selected it as the general contractor.

Together, the leading design team's (SERA Architects as the executive architect) and general contractor's (HSW) combined skills and experience ranked exceptionally high.

6.3.4.2 Team collaboration and communication

Due to local conditions, such as geographic location and owner/team issues, GSA Region 10 had been developing its own CMc protocol to accommodate and guide local practices. This development and deviation from the conventional CMc made Region 10 CMc more similar to IPD. As illustrated in Figure 6.2d, the team organization closely resembled an IPD project team. Factors unique to the EGWW team that did not always occur in an IPD team are listed below:

- GSA had an on-site manager during the construction phase
- Selection of major first-tier subcontractors prior to the design being developed
- Shared co-location facilities
- Use of BIM

This unique structure enhanced the IPD team's collaboration and communication. A GSA project executive stated, "Having on-site personnel, co-located with the team, has had a significant impact. This attribute may have had a greater impact than the contract form or type" (Cheng, 2015).

6.3.4.3 Innovation—radiant panels

To determine why innovation did not significantly impact the construction cost, the researcher further investigated the statements from the project team, including from GSA program managers. Project documents revealed the important leadership role the GSA local program manager played in leading innovation efforts, to some extent even compelling innovation upon the project team.

One example was the installation design of the radiant cooling and heating ceiling system in EGWW. Because this system was and still is relatively new to the US market, the design team expressed concerns about using the system. Besides unfamiliarity of the system, a lack of air movement was another major concern. Using a radiant system eliminates the fans found in a traditional HVAC system and, consequently, the "flowing" air from the traditional system. Although a radiant system provides a high level of thermal comfort and indoor air quality, it was predicted that occupants accustomed to noisily blown cool or warm air (as in traditional systems) would perceive a problem, linking minimal air movement to a lack of cooling. One way this was corrected during post-occupancy work was to adjust the range of temperature allowed to improve the balance between energy savings and occupant comfort (AIA Top Ten COTE). A design team member stated,

> For me, it was unique to be in a position to see that my team was expressing a significant concern and having the owner in the group say, 'We understand. We think that the benefit is worth the risk.' Having that really clean and open dialogue. That's not your everyday life [in the building industry].

In this case, the innovation was driven by the determination of the client, rather than the project designer, or at minimum, the client and project team played equally important roles in implementing an innovative solution. Furthermore, the skills and experience of the project team outweighed innovation due to the project team's conservative shift to ensure the design solution could be delivered, which differs greatly from conventional projects.

6.3.5 Construction cost

The project had a GSF of 46,122m^2, a construction cost of $141.5 million, and a normalized cost of $3,066/m^2. Two project-specific characteristics contributed to the control and reduction of the construction cost. The first was related to the project's contractual/team organization. The project budget and schedule were organized in a highly collaborative way since the project team shared the profits and risk. While the GSA owner was the final decision-maker, the contingency was treated as a pool of money used to benefit the project, and decisions for its use were shared. In particular, savings were attributed to how the value engineering of the GMP budget was inclusive of both prime and first-tier subcontractors. This collaborative decision-making and contingency-sharing mechanism was used to reduce the project budget. As a full renovation project, one of the largest risks was the unforeseen condition of the existing building. EGWW could take a team approach to decide how the contingency would be used to address those unforeseen issues because of the integrated team structure.

The second characteristic was the use of BIM. EGWW was among the earliest GSA projects to fully utilize BIM from design to construction. The project team was highly innovative in its use of BIM in developing an information room (which they dubbed the iRoom), a combination of centrally managed BIMs in a co-located office. From the early design phase, the team members, including subcontractors, had access to each discipline's design via co-located BIM models on a shared server in the iRoom. This open access within the team enabled on-the-fly coordination among disciplines. In addition, first-tier subcontractors were also co-located so they could provide a constructability review during the design phase, using a virtual model, to reduce risks and mistakes that typically happen at the construction site. This early trouble-shooting also contributed to a cost reduction. The EGWW team not only used BIM for real-time coordination but also developed a unique BIM process called BIM snapshots. Instead of structuring the delivery of drawing packages following conventional project processes (e.g., SD, DD), BIM snapshots created drawing packages by capturing images at specific moments during design, after which the project design team printed drawing sets from the BIM model. Those BIM snapshots were then aligned with the CMc's contingency strategy, made possible by back-and-forth coordination between the CMc and the architect. The BIM snapshots, together with printed drawings, were used by owners to validate the design intent and by the CMc to solicit subcontracts. This approach allowed the project team and owner to prioritize within the design process, deferring noncritical portions of the design to later phases. This helped maintain an aggressive schedule. The team estimated a $940,000 saving in project costs because of the reduction in hours spent on design documentation—from a typical schedule of 53,000 hours over 24 months to 44,000 hours over 15 months (Cheng, 2015).

BIM models were also used to run clash detection by the design team. Clash detection allowed coordination among multiple systems, including structural and mechanical systems. The BIM clash detection and coordination enabled a 38% reduction in fabrication time by using mechanical modular skid prefabrication instead of conventional on-site fabrication (AIA, 2012). Unlike WAF, the use of BIM in EGWW was well planned with specific goals and assigned responsibility; thus the financial benefits of the innovation could be tracked and quantified. However, the use of BIM in EGWW was not unique to sustainable building and did not directly contribute to sustainability. Instead, it is a universal good practice that can be applied to all buildings.

6.4 CASE 3: HOWARD M. METZENBAUM US COURTHOUSE

The Howard M. Metzenbaum US Courthouse (HMM) is five-story, 21,297m² office building located in Cleveland, Ohio (refer to Figure 6.3a). Different from the two previous cases, HMM is not an ARRA-funded project. It was built in 1910 as a courthouse and registered as a historic place in 1974. HMM has undergone several major extensions and renovations, including adding an HVAC system in the 1960s along with an upgraded electrical system. After the original tenants were relocated to the new US courthouse, HMM became vacant in 2002. The tenant space alternation and building system modernization were done to backfill

HMM with new tenants. The complete renovation was finished in 2005 with the goal of historic preservation and energy efficiency upgrades. The renovation included a complete replacement of the HVAC, electrical, fire/life safety, security, alarm, and communication systems. The project also converted the basement space occupied by US Marshals to parking. The estimated total project cost was $29,401,000 (General Services Administration, 2002), the actual total construction cost provided by GSA was $44,600,000, and the normalized unit cost was around $2,100/m² ($189/ft²). Since the final construction cost was almost double the original budget, hence, HMM was selected to understand the underlying factors and reasons for cost.

6.4.1 Technical complexity
Since the project was completed early with a lower LEED certification requirement from GSA, the technical complexity of modernizing the building system was not significant. The greater challenge was induced by the conflicting needs of renovation/modernization with the preservation of the buildings' historical features and significance. For example, after the construction team removed the drop ceilings, which were added in the 1960s to conceal the ducts for the air-conditioning system, they discovered the original ceiling and ornate plasterwork (*Howard M. Metzenbaum U.S. Courthouse*, n.d.). To reveal the original historic ceilings and ornate plasterwork, the project design team placed mechanical chases and risers into the defunct chimneys. While using a mechanical chase in the chimney was an innovative solution, it also increased the challenges in installation and maintenance, which were associated with a certain amount of cost increase. Second, a substantial portion of the additional cost beyond the original budget was solely dedicated to historic conservation and restoration because of HMM's historic place status. For example, the project reinstalled 35 restored murals by the nineteenth-century American artist Francis Davis Millet (DLG Group, n.d.), which were not included in the original budget. Furthermore, most costs associated with historic preservation or restoration were not directly related to sustainability or energy efficiency.

Regarding the building system complexity, most upgraded HVAC, electrical, and plumbing systems were the same as those used in new conventional constructions, with no substantially higher cost. However, one advanced mechanical system with an additional cost was the demand-controlled ventilation system, which connected to CO_2 sensors inside the room so the ventilation could provide higher indoor air quality by adjusting the outside air intake while preserving energy.

6.4.2 Sustainability
Compared to other included case projects, HMM had a relatively low sustainability achievement according to its LEED level. However, it was the first adaptive reuse project in the GSA portfolio to achieve LEED Certified level certification

Figure 6.3
The Howard M.
Metzenbaum US
Courthouse: (a)
after renovation
(Creative Commons
Attribution-
Share Alike
4.0 International
license), and
(b) HMM team
organization

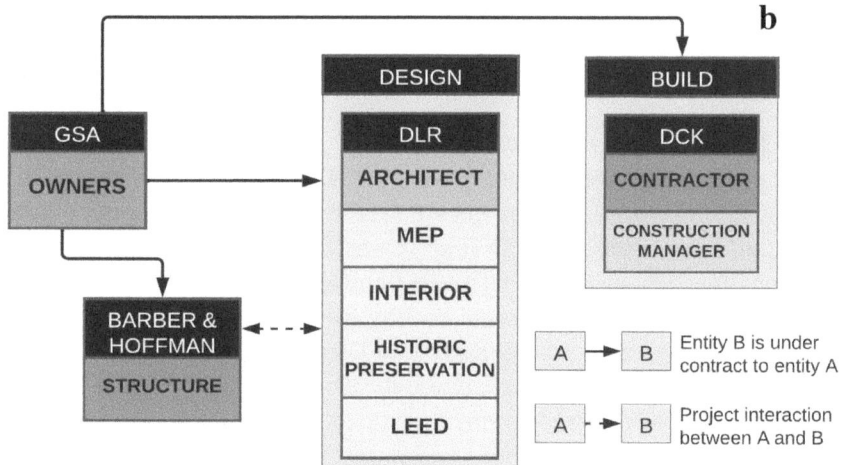

among GSA's First 40 LEED Projects. Its LEED score is 44 out of 110, with actual energy use being 84 kBtu/ft^2 and water use being 2.14 gallon/ft^2.

6.4.3 Procurement and contract

HMM used the DBB delivery method with a CMc. WRL was the leading design firm, providing MEP engineering, interior design, historic preservation, and LEED consulting services (refer to case 1). Dick Corporation (DCK) was hired as the construction manager and general contractor.

6.4.4 Project team characteristics

6.4.4.1 Team skills and experience

As introduced in the WAF case, WRL had become the leading expert in sustainable design. However, in 2005, WRL did not have much experience in LEED buildings. DCK, a Pittsburgh-based company was ranked 345th among the top 400 international contractors (DCK, 2019), albeit not within the rank of top green contractors. Compared to other higher-ranking contractors with extensive experience in sustainable construction, DCK had more strength in multiple-family units and hotels. Together, the leading design team and general contractor, had combined skills and experience that ranked average. In 2005, general knowledge of sustainable design in the architecture, engineering, and construction sector was not sufficient; consequently, it was understandable that GSA only targeted the LEED Certified level.

6.4.4.2 Team collaboration and communication

As illustrated in Figure 6.3b, the project team organization represented a conventional DBB project. The design team and contractor had limited interaction, with the main communication being between the owner (GSA) and the design team, consultants, and contractors. The author was unable to find further details of collaboration and communication among the project team; therefore, it was assumed the level of collaboration and communication was average as in other DBB projects.

6.4.4.3 Innovation—space alternation

Most design innovation focused on key space alternation rather than technological advancement (refer to Figure 6.4). The design converted the previous courtyard (open to the sky) to an atrium covered by a glass skylight. This change in space function solved the circulation problem, increased security requirements, and allowed the historical corridor systems to be allocated to active tenant use (*Howard M. Metzenbaum U.S. Courthouse*, n.d.). The project team claimed that the covered atrium, which functioned as a light well, dramatically reduced the building's energy use through lowering the demand for artificial lighting. However, to the author's knowledge, no data were found to support this claim.

6.4.5 Construction cost

The HMM project was completed in 2005 and had a total construction cost of $44,600,000 and a normalized cost of $2,100/m². When converted to net present value (2021) with a 6% inflation rate, the cost was $5,033/m², which was higher than the cost of a regular renovated building. The initial budget for this project was $29,401,000, with a normalized cost of $1,288/ft². The cost overrun was 63%, with two potential causes for such a high-cost overrun. First, the project team was inexperienced in the LEED rating system and general sustainable design, especially from the contractor and client side. Second, beyond

Figure 6.4
HMM renovated courtyard/atrium space (photo by Library of Congress)

regular renovation and system upgrade costs, most of the construction cost was related to space alteration: converting the open courtyard to a closed atrium space. The contribution of energy efficiency of the space alternation was not well documented; consequently, the cost could not be completely contributed to a sustainability pursuit.

6.5 CASE 4: FEDERAL CENTER SOUTH BUILDING 1202

The Federal Center South Building 1202 (FCS) was a new construction located in Seattle, Washington. The project began in 2010 and was completed in 2012. Built as the regional headquarters for the US Army Corps of Engineers (USACE) Northwest District, it is a three-story office building with a gross area of 209,00 ft^2 (refer to Figure 6.5a). It was funded through ARRA and shaped by a government program designed to spur innovative high-performance design for federal facilities. The FCS project was delivered on time and within budget by an integrated DB team. The project was driven by aggressive high-performance goals and a tight schedule, and the design integrates active and passive systems and material reuse in innovative ways that will be described in the following sections.

The project site is located on the bank of Duwamish Waterway, which was an industrial site with poor air quality due to a nearby cement plant. The waterway transformed a 4.6-acre brownfield site from a 100% impervious to a 50%

pervious landscape. On-site stormwater runoff is treated within stormwater surface ponds, rain gardens, and wet ponds and then filtered. At the end, 100% of the stormwater can be collected and treated on-site using natural landscape, thereby eliminating the need to connect the already overburdened city stormwater system (AIA Top Ten, 2013).

Unlike the previous three cases, FCS was a new construction, so the project team had opportunities at the early stage to use passive design strategies to optimize orientation and massing for annual energy use reduction, as well as to reduce the peak load, consequently reducing the size of the mechanical system. The building massing is north and south oriented with a central atrium opening to the west. In addition, the shading system was designed specifically with different orientations, with vertical fins and a variable number of horizontal louvers (depending on the orientation) to control solar heat gain, thus reducing the annual cooling load (refer to Figure 6.5b). The building has a relatively narrow

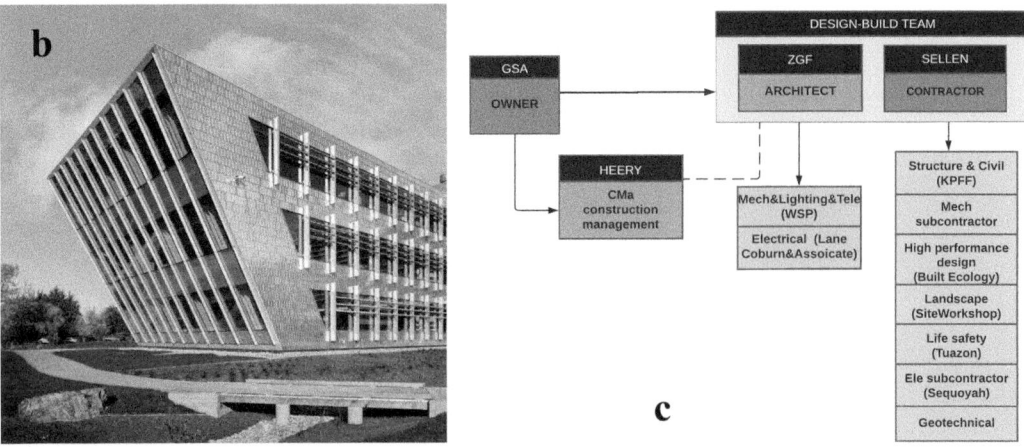

Figure 6.5
Federal Center South Building 1202: (a) main entrance, (b) west-facing façade with louver, and (c) team organization

floor plate of 18.2 m, which allows for daylight partial penetration (WBDG, 2017). The project achieved LEED Gold certification, with a score of 61/110. The total construction cost was around $73 million, and the unit cost was $3,888/m^2.

6.5.1 Technical complexity

6.5.1.1 Floor system

Due to the nearby cement plant creating air pollution, natural ventilation was not feasible because of particulates in the air and Department of Defense security requirements. A 100% filtered outdoor mechanical ventilation was used, and the air is delivered underfloor (AIA Top Ten, 2013). The first reason for choosing underfloor air distribution (UFAD) is mainly for its advantages in energy efficiency, since the UFAD decouples ventilation from the heating and cooling system. Second, UFAD can provide the necessary space for electrical and IT distribution. However, the building design has a curved floor plan (refer to Figure 6.6b), which required many floor panels to be custom cut and increased the number of joints that needed to be sealed (Chaloeigheep et al., 2014; WBDG, 2017). Consequently, the unique building shape accumulated certain additional costs.

6.5.1.2 Structural system

As shown in Figure 6.6c, a diagrid structure wraps around the exterior of the building, which meets GSA's security requirements for progressive collapse. The structure is essentially an oversized truss comprised of diagonal columns and bracing that function to resist both lateral and gravity loads (WBDG, 2017). Because of the unique building shape, many structural elements needed to be custom designed and manufactured.

6.5.1.3 HVAC system

FCS's mechanical system is unique in three perspectives (refer to Figure 6.6a). First, to utilize both daily and seasonal heat exchange patterns to collect and store thermal energy for future needs, a thermal storage tank containing phase change materials was built. During the morning, the thermal storage tank freezes at 55°F. The building uses the free cooling generated by the heat recovery chillers. Then, the frozen thermal tank is later used to cool the building at midday (AIA Top Ten, 2013). However, this new technology did not fully perform as expected. The shortcomings of the thermal tank included a lack of density of phase change materials and insufficient flow rates, which can be overcome in future projects by increasing density and flow rates (Chaloeigheep et al., 2014).

Second, this project was one of the first in the region to use structural piles for geothermal heating and cooling with a heat pump. Similar to the thermal storage tank, the ground loop system cycles the free heat or cooling from the ground when there is not enough storage in the thermal tank (Cheng, 2015). The success of the energy piles can be attributed to the high rate of exchange of ground water found adjacent to the site. In general, energy piles are a viable option when depth to bearing soil is significant (Chaloeigheep et al., 2014). The

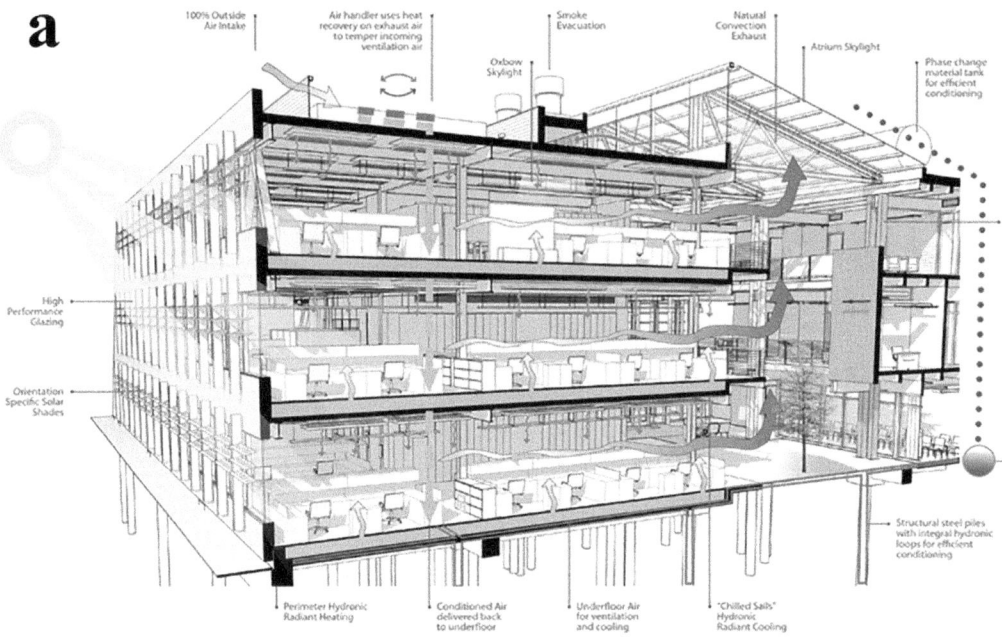

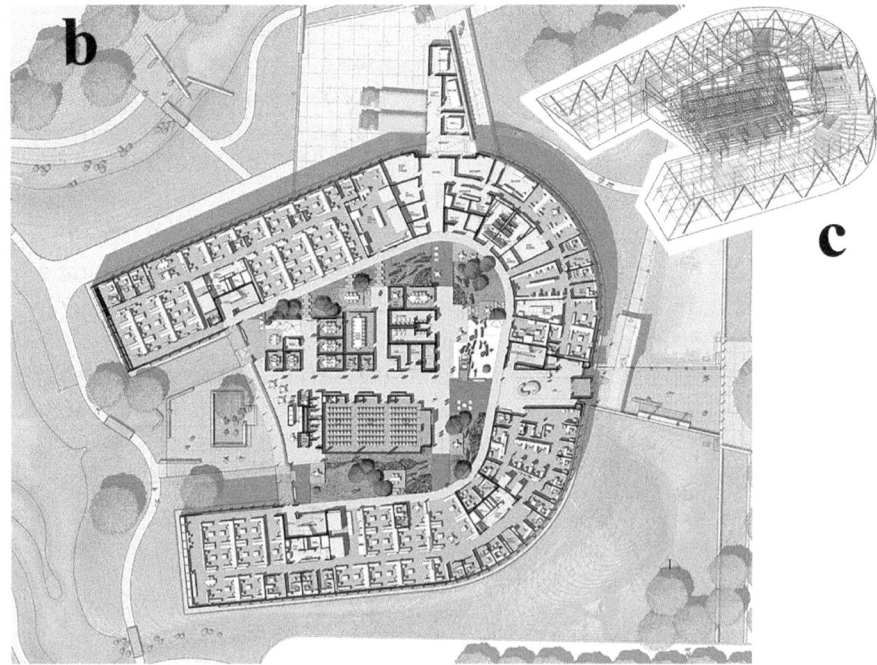

Figure 6.6
Federal Center South Building 1202: (a) section perspective, (b) ground floor plan, and (c) structural diagram (photos by GSA)

diminishment of effectiveness of the thermal storage tank was made up by the geothermal loop.

Third, a new product, a chilled sail ceiling system was developed and custom manufactured specifically for this project to achieve the energy efficiency goal. Supplemental cooling is delivered through an energy-efficient system of passive chilled sails suspended from the ceiling. Chilled sail technology was developed in Europe in the late 1990s but is still relatively new in North America. It couples with the radiant cooling effects of standard radiant panels with a convective component in cooling for increased thermal comfort. The sails' unique shape gives them more surface area than a traditional radiant panel, increasing their radiant capacity while still achieving the high comfort of radiant systems. They allow air that has been cooled by contact with the sails to pass through openings between the blades, thus increasing the capacity of the unit and providing an effective means of dealing with the sensible cooling load (Rimmer, 2013).

6.5.2 Sustainability

The energy model predicted the building would operate with an EUI as low as 20.3 kBtu/ft^2/yr, performing 40% better than ASHRAE 2007 (AIA Top Ten, 2013). Measuring the performance of the building began in January 2013, less than three months after substantial completion of the project. The first few months of measurement and verification (M&V) required coordination between the designers, builders, controls contractor and building operator, and the entire M&V team. During the first quarter of operation, the M&V team tuned the building's HVAC systems, which were running beyond the scheduled operation hours, and also improved indoor thermal comfort and acoustic comfort. After one year of operation, the building was tuned from operating at 10% over the target in the first quarter to 12% below the target in the fourth quarter. The final first-year EUI was 25.7 kBtu/ft^2/yr (adjusted for plug loads and operation outside the operational hours) (Chaloeicheep et al., 2014).

6.5.3 Procurement and contract

FCS used a DB project delivery with a fixed-price contract. The GSA also included a performance clause in the contract and a post-occupancy verification phase. The performance clause and verification phase were developed by GSA's Northwest/Arctic Region office and used for the first time in this project (Cheng, 2015).

The DB team comprised the architect (ZGF Architects LLP) and contractor (Sellen Construction). The procurement followed a two-step procurement process, combining qualification-based selection and best value selection. The process began with a three-month design competition wherein all competing teams proactively proposed a best value solution, a schematic design-level proposal. Three teams participated in the competition unfunded, each taking a substantial risk to secure a rare, fully funded project at the end of the Great Recession (Chaloeigheep et al., 2014). This resulted in a proposal to GSA that based design decisions on cost-effective, high-performance design. Finally, the winning team guaranteed GSA a project that fulfilled the requirements for sustainability metrics, the schedule, and the budget. In addition, the DB team was required to include a full list of consultants in the proposal. Therefore, Sellen Construction and ZGF Architects quickly worked with GSA to finalize consultant selections as show in Figure 6.5c.

The fixed-price contract was unique in the way it gave the contractor maximum risk and full responsibility for all costs resulting in profit or loss. To a certain level, the contractor functioned as an owner of the construction project. It was Sellen Construction's first experience with this type of contractual structure, and its senior vice president stated,

> I was in charge of spending this money efficiently. As a contractor, usually, you can make recommendations and suggestions, but it's the owner who usually has final decision power. In this case, I was in the position to make those decisions. I have to say the role is harder than normal contracting. It is much more intense.
>
> (Cheng, 2015)

6.5.4 *Project team characteristics*

6.5.4.1 Team skills and experience

ZGF Architects was found in 1942 in Portland, Oregon, and it ranked fifth in *Building Design+Construction's* Top 160 Architectural Firms for 2021 and 77 in ENR's 2021 Top 500 Design Firms. It also ranked 14th in ENR's 2020 Top 100 Green Buildings Design Firms. Sellen Construction was founded in 1944 in Seattle, Washington, and is now one of the area's most prominent construction companies. It ranked 165th in ENR's 2021 Top 400 Contractors and 49th in ENR's 2021 Top 100 Green Building Contractors. ZGF and Sellen had more than 20 years of experience working together in settings other than DB, had worked on projects together in the recent past (Cheng, 2015), and had both originated in the Pacific Northwest region. Their knowledge and experience of construction in this region contributed significantly to their competitiveness among other DB team competitors.

6.5.4.2 Team collaboration and communication

FCS used a DB contract with a CMa. As illustrated in Figure 6.5c, the main collaboration and communication occurred between the DB team and GSA. Most consultants were supervised by Sellen and worked with ZGF. A strong collaborative culture emerged during the proposal phase, which was marked by a strong mission alignment and mutual trust and respect. The ZGF team perceived Sellen Construction as "allow[ing] the design team [to have] a little more rope than a lot of other contractors do, allowing them to explore ideas and have enough time to let good ideas really come to the top" (Cheng, 2015).

An important contractual factor, the performance clause, reinforced the collaboration and communication of the DB team. GSA included performance clauses in the contract as a mechanism to clearly communicate the project goals, aligning the team with the owners' priorities. The building had one year to meet the energy performance targets; until then, 0.5% of the original contract award was withheld from the DB team. Consequently, the DB team understood that falling short of their energy performance target meant the team would not receive the full compensation, which motivated them to work diligently through close collaboration (Cheng, 2015).

6.5.4.3 Innovation

The biggest innovation in this project was the use of reclaimed wood to reduce construction waste and embodied carbon. The DB team reclaimed approximately 200,000 board ft of structural timber and 100,000 board ft of decking from the decommissioned non-historic World War II warehouse on-site, which resulted in a 99% diversion rate of construction waste. Using a phased demolition process, wood components were individually harvested from the warehouse. The team pulled nails, unfastened bolts, removed brackets and devices, trimmed out fractures, and sorted the wood before it was shipped to a local mill for structural grading and fabrication for use in the new building (Chaloeigheep et al., 2014).

To optimize the use of structural timber, the DB team used a composite design for the floor system. The team believed they were the first to use this composite timber/concrete system in the United States, which reduced structural material needs by 30% (AIA Top Ten, 2013). Being the first time this design type was used nationally, the team built a mock-up in the adjacent warehouse to test structural strength and verify it met the code requirements. Testing the new and customized assemblies required close collaboration among architects, engineers, fabricators, and contractors. In the warehouse, Sellen Construction constructed the proposed assembly, and the structural engineer (KPFF) installed sensors to monitor the movement of the assemblies under the lateral load. The testing results of the mock-ups demonstrated the proposed assemblies exceeded the standards; thus the DB could use this innovative structural system using reclaimed wood (Chaloeigheep et al., 2014).

6.5.5 Construction cost

FCS had a total construction cost of $73,314,341, a normalized cost of $3,888/m^2, which was much less than a typical newly constructed office building in the Pacific Northwest – $5,000/m^2 at that time (Chaloeigheep et al., 2014). FCS was completed in 2012, and when converted to net present value (2021) with a 6% inflation rate, the cost was $5,233/m^2, still lower than building a new conventional mid-rise office building, at $6,188/ft^2.

The success of controlling the budget was made possible by two mechanisms: the transparent contingency provided by GSA and a "betterments" list developed by the DB team. GSA as the final decision-maker treated the contingency as a pool of money to be shared by the entire team to benefit the project, so the value of the contingency was known by the entire team. The "betterments" list described a wish list of items not originally included in the budget, which the team believed would improve the building's energy performance and design excellence. Those items were implemented when funds (transparent contingency) became available, whether through savings, release of contingency, or additional funds. Items that were implemented included using fritted laminated glass in the skylight to optimize daylight, installing geothermal piles, upgrading lighting controls, adding occupancy sensors at workstation task lights, and building a rainwater harvest system.

6.6 CASE 5: BYRON G. ROGERS FEDERAL BUILDING AND US COURTHOUSE

The Byron G. Rogers Federal Building and US Courthouse (BRF) was built in the 1960s in downtown Denver, Colorado. The history of this building complex originated in the years following World War II, when the population of Denver grew rapidly as numerous federal agencies relocated to the city. The existing federal building could no longer accommodate growing space needs, and the government began planning for a new complex. In 1961, GSA purchased the downtown Denver site, the groundbreaking occurred the same year, and the first occupants moved into the building in 1965. As illustrated in Figure 6.7a, the complex is com-

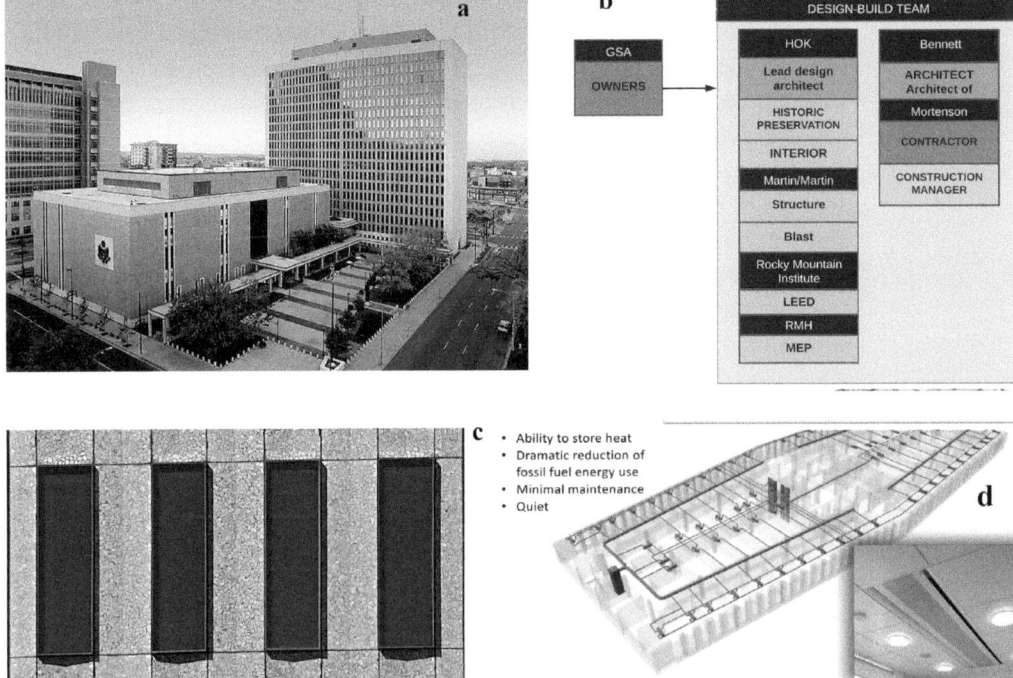

Figure 6.7
Byron G. Rogers Federal Office Building and US Courthouse: (a) after renovation, (b) BRF project team organization, (c) façade, and (d) BRF chilled beam mechanical system (photos by GSA)

posed of three parts: an 18-story office building with a gross area of 44,460 m², a five-story courthouse with a gross area of 22,230 m², and a shared landscape plaza. In October 2016, BRF was listed on the National Register of Historic Places.

Regarding the building's historical significance in design, the building facade and site plaza were notable representations of mid-twentieth-century formalist style architecture (GSA, n.d.). The lower portion of the façade was cladded with marble panels and the upper portion had precast panels, which were unique design features embodying the building's historical significance. This unique façade design in fact created difficulty for the renovation, which will be discussed in later sections. The courthouse renovation was completed in 2006 (cost of $52.3 million), and the office building followed in 2013. The office renovation project achieved LEED Gold certification, with a score of 66/110. The total construction cost was around $194 million, and the unit cost was $3,477/m².

6.6.1 Technical complexity

The primary goal was to upgrade all the major buildings systems. This included the replacement of mechanical, electrical, and plumbing systems, and the interior of the building was almost completely gutted, except for the historically significant interior elements, the primary structure, and the historic exterior envelope. This complete renovation provided the design team with an opportunity to have a systematic approach and a more efficient and flexible arrangement. For example, the mechanical and electrical rooms were moved to building ends, freeing up room within the central core to better serve the different mechanical zones of the building (Bartels & Swanson, 2016).

6.6.1.1 HVAC system retrofit: active chilled beam

The new mechanical system was an active chilled beam (ACB) combined with a thermal storage system that transfers heat from the warmer side of the building to the cooler side and vice versa (refer to Figure 6.7d). This system allows the building to store unused energy and utilize it in areas that need it the most through the day. A regular chilled beam system primarily uses moderate temperature water to condition the building space without a thermal storage system. A unique characteristic of ACB is the combination of a hybrid magnetic bearing heat recovery chiller and a 50,000-gallon thermal storage tank that is designed to circulate captured heat (from occupants, computers, lighting, etc.) throughout the building. The tank was sized to accommodate the heating and cooling needs overnight and on weekends by only operating pumps and stepping down chillers and boilers during unoccupied hours (Bartels & Swanson, 2016).

Compared to a conventional air-based mechanical system, the ACB system presented an energy use reduction opportunity because it provides a method for transferring energy between different zones within the building efficiently so heat and cool air will not be lost. The primary energy saving comes from a reduction in required cooling – an active chilled beam uses 58°F (14°C) water for cooling as compared to 45°F (7.2°C) water for a typical air-based cooling system. The heat mode is also more efficient in a water-based system (Tupper et al., 2012).

Another advantage of using a water-based chilled beam system is that it requires significantly less space than a conventional air-conditioning system (with large ducts). The small dimensions of water pipes in the system allow for greater floor-to-floor heights. With greater ceiling heights and the removal of induction units, 30% more window area is created for daylight and views (Bartels & Swanson, 2016).

6.6.1.2 Lighting system

This project had an ambitious goal to reduce the lighting power density to 0.55 W/ft^2 (compared to the code-required 0.9 W/ft^2) using LED lighting fixtures. Commercially available LED lights have become common practice now; however, at the time of the building's renovation, the standard office-type LED fixtures were not available. Therefore, the project team organized a design competition to compare custom-designed LED fixture prototypes from different manufacturers. From this competition, a practical and affordable lighting fixture prototype was selected, which then went into full-scale production (Bartels & Swanson, 2016).

6.6.1.3 Building facade

Renovating the building's façade system may have been one of the most challenging tasks because of the historic preservation requirement. Four primary aims that were addressed included (a) maintaining the historic building design, (b) reducing the heat gain through increasing the thermal property of the facade and maintaining vapor barrier continuity around existing structural elements, (c) strengthening the existing structure to meet the current code by installation of blast windows and supporting structures, and (d) maximizing daylight.

To maintain the historic design, challenges involved keeping the existing precast panels in place while retrofitting the windows. A prominent historic design feature of the façade was a pattern of slender aluminum windows (refer to Figure 6.7c), that were projected out from the precast panels on the upper floors. These design features created two challenges. First, the window frame design created a critical thermal leak since the aluminum's high heat conductivity transferred heat in and out of the building, which greatly affected energy efficiency. Second, the existing precast panels were irreplicable and known to be brittle when handled (Bartels & Swanson, 2016). An innovative solution was to maintain the existing precast panels and exterior windows and install a new blast window on the interior, so all work could be performed from the building's interior without damaging the existing exterior façade (refer to Figure 6.7c).

Besides adding high-performance windows, additional insulation was added to the exterior walls and roof. Before the renovation, the roof's R-value was around R-20, and the exterior walls were estimated at R-5. Following the renovation, the exterior walls achieved a value of R-20, which is about four times higher.

6.6.2 Sustainability

BRF achieved LEED Gold certification and earned an ENERGY STAR rating of 99. Before the renovation, the building consumed 91.8 kBtu/ft²/yr, close to the national average of 93 kBtu/ft²/yr (Tupper et al., 2012). Following the renovation, there was a 68% reduction in building energy use and an EUI of 38.4 kBtu/ft²/yr, which was 50% better than ASHRAE 90.1 2007 (Mørck et al., 2017). Daylighting increased by about 25% from the existing condition. According to GSA, the renovation provided a carbon emissions reduction of approximately 2,908 tons (Apogee Renovation, 2019).

6.6.3 Procurement and contract

This project used a DB delivery method, and the team was selected based on its qualifications and best value. The DB team comprised Mortenson as the design-builder, HOK as the design architect, Bennett Wagner & Grody Architects as the architect of record; Martin/Martin as the structural engineer and blast consultant; the RMH Group as the mechanical, electrical, and plumbing engineer consult; and Rocky Mountain Institute as the high-performance green building consultant. Team members had long-term relations with each other.

6.6.4 Project team characteristics

6.6.4.1 Team skills and experience

HOK was a leading design firm, ranking fifth in *Architectural Record's* Top 300 US Architecture Firms of 2021 and fourth in ENR's 2021 Top 100 Green Buildings Design Firms. Bennett Wagner & Grody Architects, now part of CannonDesign, ranked 13th in ENR's 2021 Top 100 Green Buildings Design Firms. The construction manager and general contractor was Mortenson, a Minneapolis-based company, which ranked 16th in ENR's 2020 Top 400 International Contractors and 20th in ENR's 2020 Top 100 Green Building Contractors. Together, the leading design team and general contractor had combined skills and experience that ranked higher than average.

6.6.4.2 Team collaboration and communication

As illustrated in Figure 6.7b, the project team's organization reflects a DB project. The main communication was between the owner (GSA) and DB team. As indicated by a senior project manager at Harmon, Inc (the glazing contractor for the project, hired and supervised by Mortenson), "The project was a success due to the collaborative efforts by the entire design and construction team." For example, Mortenson and Harmon enjoyed a long-term relationship and had collaborated on dozens of projects, several of which used Wausau's windows and curtain wall systems. Wausau had also contributed to many HOK-designed projects.

6.6.4.3 Innovation: technology employment and behavior change

To maximize sustainability for a reasonable price, the DB was innovative in optimizing strategies to reduce the energy load through a balanced design maximizing daylight while reducing heat loss through large windows. To boost daylighting, the team removed the existing induction units and raised the acoustical ceiling heights, which were installed during previous renovations – the ceilings and induction units had obstructed 30% of the windows (Mørck et al., 2017). Where open floor plans were not possible because tenants valued private offices, the team considered translucent or transparent walls for private offices. The original large windows had a high potential for heat loss; therefore, to minimize heat loss and thermal leakage, all windows were fitted with high-efficiency glazing with a thermal break. In addition, as described in Section 6.6.1, additional interior glazing was added to the upper floors where replacing the existing windows was difficult. This was a unique and innovative retrofit solution that can be applied to other historic buildings with high-energy performance goals.

In large office buildings, plug and process loads (PPLs) contribute significantly to total energy use, around 47% (Department of Energy, n.d.). PPLs are defined as the loads in a building that are not associated with heating, ventilating and air-conditioning, lighting, water heating, and other major equipment needed for basic building operations (Lobato et al., 2012). PPLs refer to the loads from computers, monitors, printers, refrigerators, water purifiers, and other office devices. The percentage of total building use from PPLs is increasing: by 2029, it is anticipated to increase to 51% (EIA, 2022). Reducing PPLs requires technological advancements as well as behavior changes. The BRF project team engaged with tenants early on and published a sustainability guide that included information about reducing plug loads through using efficient appliances and shutting off devices when not in use (Mørck et al., 2017).

Table 6.2 BRF cost breakdown of major building system renovations (Mørck et al., 2017)

Items	Description	Cost
Mechanical system	Active chilled beam system Dedicated outside air system Chiller, cooling tower, and pumps Thermal storage Others	**$15,500,000**
Lighting system	LED fixtures Daylight sensors and dimming controls Occupancy sensors Others	**$4,200,000**
Glazing system	Reglaze existing frames with 1-inch insulation glazing interior dual pane blast window 1/4 inch crystal gray w/VE-2M #2 Assembly U-0.14, SC-0.27, Tvis: 37%	**$4,650,000**
Roof and walls		**$815,000**
Plumbing fixture	Low flow fixture	**$447,000**
Total	Major building system renovation	**$25,612,000**

6.6.5 Construction cost

Completed in 2013, the BRF project had a total construction cost of $194,295,701 and a normalized cost of $3,477/m². When converted to net present value (2021) with a 5% inflation rate, the cost is $3,722/m², which is lower than renovating a conventional government building, at $4,733–9,377/m² (Smart Remodeling LLC, 2022). As listed in Table 6.2, the total spending on major building system renovations that contributed to the building's energy efficiency was approximately $25,612,000, around 13% of the total construction cost. The largest item, a full mechanical system renovation, comprised 8% of the total construction cost. For a conventional building, although there is no fixed percentage for the cost of major building systems, the mechanical system normally contributes 4.4% of the cost, with the plumbing system comprising 4.3% and the electrical system around 4.2%. All together, they total 12.9% of the construction cost (Gerardi, 2021), similar to the BRF percentage. However, the percentage of BRF's mechanical system cost was much higher, at 8% versus 4.4%, clearly relating to several advanced and costly building systems that are not used in a conventional building, such as thermal storage and an active chilled beam system.

6.7 CASE 6: GEORGE THOMAS "MICKEY" LELAND FEDERAL BUILDING

The George Thomas "Mickey" Leland Federal Building (GTML) was built in 1983 in downtown Houston, Texas. It was originally built as a 22-story office building with a detached six-story parking garage (refer to Figure 6.8). The total gross area was around 32,940 m². After GSA purchased the building in 1987, an energy performance analysis documented that air and water infiltration could be mitigated through building modernization. When ARRA funding became available in late 2009, a request for a DB proposal was issued with a scope including a complete façade replacement to address the air and water infiltration issue and energy conservation, and an upgraded structural system to meet a wind load requirement that had recently been changed in the building code (Cheng, 2018, p. 2). The renovation began in 2010 and was completed in 2015.

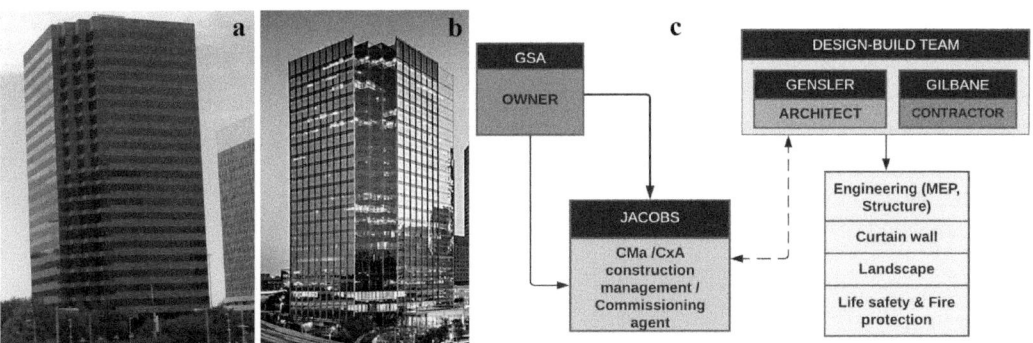

Figure 6.8
George Thomas "Mickey" Leland Federal Building: (a) before renovation, (b) after renovation, (File: Mickey Leland Federal Building.jpg from Wikimedia Commons) and (c) team organization

6.7.1 Technical complexity

6.7.1.1 Imperfect existing condition: building facade recladding

A foreseeable challenge for high-rise building façade recladding is the imperfections that occur in the original construction. The request for proposal for DB services indicated the building would be vertical and true; however, the project team scanned the building twice and discovered the building was significantly out of plumb and twisted and some elements exceeded the property line. Because of the imperfect geometry, the DB team had to redesign the exterior curtain wall quickly to accommodate the various dimensions and twisted conditions. The redesign and redocumentation incurred costs to both the DB team and client (GSA) and increased the total construction cost. Regarding the property line encroachment, GSA negotiated with the city of Houston to obtain an encroachment easement and extend the building down to the public right of way, without extra costs (Cheng, 2015).

6.7.1.2 Construction logistics

One challenge of the project was that the building remained fully occupied during construction and the team had to work around unknown conditions while keeping the building operational. An example was managing the sawtooth corners' exterior façade renovation (refer to Figure 6.8). Although the project team anticipated a difficult construction sequence, it was more complicated than expected. Because of the unique geometry at the corners, additional safety supports were needed for workers as well as multiple steps to mount the structure for the new curtain wall system. The originally planned construction sequence involved the work crews handling two floors, completing the work, and then moving on to another two floors with the tenants being temporarily located to swing space during the renovation. Regarding the sawtooth corners' condition, because of the additional structure supports, to accommodate the moving of tenants, the curtain wall construction would have to be discontinuous, which would have hindered progress as skilled welders cycled off the project. At the time of the project, there was a scarcity of welders, which meant once the welders moved to different projects during the pause period, it would have been difficult for them to return when the work resumed (Cheng, 2015). An alternative solution emerged when the tenant group occupying the sawtooth corners informed GSA and the project team that they were considering reducing their space to a smaller footprint. Having some corner space unoccupied would allow the team to schedule work in those areas independent of the main floors and keep the welders working continuously. Without the tenant group's accommodation, the façade renovation progress would have been severely impacted as well as the construction schedule and cost.

6.7.2 Sustainability

While GSA originally intended for GTML to achieve LEED Silver, the early submittal of the design indicated the potential to achieve LEED Gold. By the end of construction, the project was on track to achieve LEED Platinum, with an approximate 45% reduction in energy use (Cheng, 2015). The old building's EUI was 54.6, and compared to a typical Houston office building (EUI of 90), it performed better. After the renovation, the EUI was reduced to 31.4 (Cheng, 2015). The project was certified as LEED Platinum in 2016 and obtained a score of 85/110.

6.7.3 Procurement and contract

GTML was a DB contract with a CMa. This project initially chose a CMc as the project delivery method prior to receiving ARRA funding. Because ARRA required much of the funding to be spent as quickly as possible (its goal was an economic stimulus after the 2008 recession), GSA decided to use a DB method instead of CMc to help streamline and simplify the logistics and coordination.

Many of the team members were selected based on prior relationships between firms or between individuals within firms and GSA (Cheng, 2015). Gilbane and Gensler were hired together as the DB team; they had worked on multiple DB projects, including the Hess Tower, a 29-story office building in Houston. Gilbane and Gensler were aware of the project a year before the official request for proposal was issued; thus they were prepared. Consequently, right after the team was selected, Gilbane chose MEP and curtain wall subcontractors quickly, which allowed the team to be integrated early in the process.

Meanwhile, Jacob Construction was hired as the construction manager in parallel to the selection of the DB team. GSA contracted Jacob Construction based on its technical capabilities; in addition, Jacob Construction was also the commissioning agent.

6.7.4 Project team characteristics

6.7.4.1 Team skills and experience

Gensler ranked first in ENR's Green Buildings Design Firms from 2016 to 2021 and first in *Architectural Record's* Top 300 US Architecture Firms from 2012 to present. The contractor, Gilbane, ranked 10th in 2021 and 11th in ENR's 2020 Top 400 Contractors as well as fifth in 2020 and fourth in 2021 in ENR's Top 100 Green Building Contractors. Together, the DB team was comprised of leading experts in sustainable design and construction.

6.7.4.2 Team collaboration and communication

As illustrated in Figure 6.8c, the project team organization was a DB structure. The main communication was between GSA and the DB team and between GSA and the construction management agent, Jacob Construction. All specialty consultants were included in the DB team, under the leadership of Gensler and Gilbane.

The prior working relationships between Gensler and Gilbane helped establish effective communication and collaboration through understanding their partner's working process, capacity, and measure of success. As pointed out by a Gilbane executive,

> We already had a good working relationship [with Gensler] (from working on the Hess building), and transitioning into the next project, it was helpful to have that basis of trust already established. . . . You knew how your teammates were motivated, what they needed to be successful, and they knew what you needed to be successful.
>
> (Cheng, 2015)

Because of prior collaboration experience and mutual trust, the team's communication within the team and with existing building tenants was effective and created pathways for solving construction challenges such as the sawtooth concern.

Besides the prior collaboration between the firms, another key factor for team building and creating a collaborative culture was the appropriate designation of positions in both the DB team and GSA. Team members noted that there were experienced leaders across the project, including at GSA, Jacob Construction, Gilbane, and the main subcontractors. The Gilbane project manager observed, "At the end of the day, when you ask how come a project is successful, it really comes down to the people involved.... That's what made this project feel special..." (Cheng, 2015).

6.7.4.3 Innovation—structural design

A high-rise building design has a unique set of challenges that do not exist in a low-rise building, with a higher wind load being one of them. The original building was built in the early 1980s, and since the initial construction, a new wind load provision had been added to the code. Additionally, the new building façade has a large "wind sail" area that required that the existing lateral load system to be checked for compliance with current requirements. The initial static analysis of the existing lateral force system showed many existing structural members were significantly overstressed and did not meet the current code requirements with the new cladding system. Instead of strengthening all structural members, the structural engineer firm Walter P Moore used an innovative analysis approach to better model the building's structural behavior. This performance-based analysis accounted for the non-linear behavior of materials by using state-of-the-art methods as opposed to the approximate methods used in conventional industry practices. According to Walter P Moore, this analysis had been used in seismic design and had never before been applied to wind loads (Walter P Moore, 2014).

This alternative structural design approach reduced the quantity of structural members required for strengthening and reduced the costs for the demolition and new materials (Cheng, 2015). The savings were approximately 1,500 tons of concrete, 175 tons of reinforcing steel, and 350 tons of cradle-to-grave CO_2 emissions that would have been generated from producing the originally intended quantity of structural metals. The design team applied for and achieved a LEED Innovation in Design (ID) credit for demonstrating that their approach reduced material quantities by 30% in the lateral load resisting system (Walter P Moore, 2014).

6.7.5 Construction cost

Completed in 2012, GTML had a total construction cost of $73,314,341 and a normalized cost of $3,888/m². When converted to net present value (2021) with a 6% inflation rate, the cost was $6,200/m², which is lower than building a new conventional high-rise office building, at $7,322/ft², and within the range of a regular renovation project, at $4,777–11,000/m² (Smart Remodeling LLC, 2022).

The documentation of the existing building was incomplete, and many of the building systems had been inaccurately documented. Besides the twisted building geometry, another unforeseen problem was the original security cameras, which were intended to be reused but were found to be outdated and not in compliance with existing requirements (Cheng, 2015). Replacing them with new cameras would have increased the construction cost; however, the project team discovered that the smart diffuser system (for HVAC and lighting) could incorporate security, thus avoiding the additional cost and project time of replacing the cameras. Without the skilled and experienced team members and close collaboration, these cost savings would not have been possible. An architect stated, "[It was] really nice that the team was able to, through the process of collaboration, come up with a better way to give them a better system very economically" (Cheng, 2015).

6.8 DISCUSSION AND TAKEAWAY

Three important takeaways can be concluded from the in-depth case studies: (a) the significance of the project team's skills and experience were reflected in the project's success, measured by the achieved level of sustainability while keeping the total construction cost within the budget; (b) collaboration and communication were predetermined by the procurement process and project delivery method; thus they were not identified as primary factors through the statistical analysis; and (c) in studied cases, the technical complexity was not always derived from the challenge of meeting energy performance—several were related to the historic preservation goal while others were related to imperfections of the existing building's physical condition. Most of the design and construction difficulties were not directly related to nor did they contribute to energy efficiency or sustainability.

6.8.1 Team skills and experience were key

For projects with a higher sustainability level and relatively low cost, "**buying a high-performance team**," instead of buying a high-performance building, was GSA's approach from the beginning. As pointed out in the 2014 GSA report on net-zero energy projects, assembling a team of dedicated and qualified members assists in attaining the established targets, for both energy performance and cost budget targets. The project team skills and experience score for the studied projects are listed in Table 6.3.

WRL was the leading design firm for case 1; it provided architectural design, interior design, MEP engineering, historic preservation, and LEED consultant services for the project. WRL was a firm with expertise in historic preservation and a strong focus on research-based design approaches. Furthermore, it was ranked nationally among the top 10 firms by peer institutions. For case 2, leading design firm SERA Architects was ranked nationally as one of the top three green design firms; this recognition is reflected in their skills and experience in sustainable building design and construction. Cases 4–6 had ZGF, HOK, and Gensler as the leading design firms who not only ranked high in experience but were also recognized as leading firms in sustainable building design. The interdisciplinary nature of these firms with experts in architecture, mechanical engineering, façade engineering, and other fields had already helped to establish and cultivate the interdisciplinary firm culture. Consequently, the project teams in these firms were quite familiar with high-level collaboration on high-performance building design.

A high-performance team can assure the delivery a high-performance product; therefore, the planning process is critical for finding the appropriate

Table 6.3 Comparison of case projects' project team characteristics

Project name	Procure-ment	DB team	Skills/ experience	Communi-cation/col-laboration	Innovation
WAF	DB	WRL/Beck	Specialist/ experienced	Above average	Geothermal system/building envelope
EGGW	CMc/GMP	SERA/Cutler Anderson/ HSW	Moderate skills/ average	Above average	Radiant ceiling
HMM	DBB with CMc	DLR/DCK	Moderate skills/ average	Average	Space alternation
FCS	DB	ZGF/Sellen	Specialist/ experienced	Excellent	Use of reclaimed wood
BRF	DB	HOK/ Mortensen	Leading specialist/ experienced	Excellent	Behavior changes and technology employment
GTLB	DB with CMa	Gensler/ Gilbane	Leading specialist/ experienced	Excellent	Structural design

people and for team building. For example, the WAF building in case 1 spent four months (February–June 2010) in the procurement phase and 15 months on design (June 2010–September 2011). The procurement phase was compressed, and GSA decided early on to use a DB project delivery method to renovate the historic building. Consequently, GSA intentionally maintained an interactive procurement process by inviting open dialogue with participating firms on how to best meet the net-zero energy goal (Cheng et al., 2014). As described by a DB team member,

> With design-build, teams have to do a lot of work at the front end to even compete. DB teams that bring proposals to the GSA need to formulate a design that is progressed far enough along in terms of infrastructure, architecture, and cost.

This level of design development in the early planning stage largely hinges on the skills and experience of the DB team. The skills and experience of the project team helped the team develop practical design solutions within the budget that were both realistic and highly innovative.

The findings of the importance of a high-performance team validates the concept of the effort curve proposed by Paulson and further developed by MacLeamy (refer to Chapter 1). As Paulson pointed out, the decisions and commitments made during the early phases of a project by the **design team have great influence** on later expenditures. This is supported by the high coefficient of soft costs that indicate the design team's input (refer to Chapter 5 data analysis results). Due to the specific funding mechanism, GSA projects have a constrained budget and schedule. To ensure project delivery at the required LEED certification level, the owner paid careful attention during the project planning and procurement stage. GSA did not use a low-bid method and also did not try to decrease the design costs, instead focusing on best value and the team's performance. This supports Paulson's second point, that any efforts to suboptimize design costs by requiring **competitive bidding for professional services** are likely to produce much **higher project costs** in the long run. The tight budget control through a high-performance team made the SBCC (even net-zero building) comparable to that of conventional building.

6.8.2 Collaboration and communication

The team collaboration method and quality in all studied projects were largely defined by the procurement and project delivery methods. The benefit of DB is not to reduce coordination but to create an integrated infrastructure. Compared to a typical DBB project, a DB process requires the project team to develop a robust design, including structural and mechanical system concepts, a budget and schedule, and an architectural concept to compete for the job during the procurement phase. By the time the GSA makes the selection, the selected team has already had more experience in design development than in a traditional DBB project. **By having more robust design informa-tion, the risk and miscommunication typically bore by the project team in the construction phase can be mitigated during the design phase**. In addition, the cost overrun burden previously carried by the contractor can be shared by the entire team. Different from a conventional project, this risk mitigation and risk-sharing infrastructure derived from the procurement and project delivery method used in a DB project typically elevates the heavy dependency of communication and collaboration among team members in a traditional DBB project. This can explain why communication and collabo-ration showed less influence from the data analysis in Chapter 5. This also partially explains why some complicated projects with a high energy effi-ciency goal (net-zero energy) were able to build within budget, because from the early stages, the project team was in agreement regarding the project goal, budget, and scope of work; thus there were minimal misunderstandings that could largely impact the construction cost later on.

It is important to clarify that the author is not trying to diminish the impor-tance of communication but instead helping readers shift emphasis from **communication quantity to quality**. With more and more convenient, advanced communication technologies and methods, the quantity of communication has increased exponentially (e.g., text, emails), but the quality of communication has not seen matchable improvements. An information overload can play a counter-productive role.

The communication frequency and volume can be less in DB than that in DBB due to the limited communication channels, which can be interpreted as less communication. In a DBB project, there are many communication channels between design firms and between firms and clients; thus overcommunication can cause miscommunication and information overflow. Conversely, with con-trolled communication channels and a well-organized communication method, in a DB project like case 1 or a CMc project like case 2, effective communication and collaboration can help the team overcome technical and budget challenges. Additionally, communication and collaboration are also determined by team skills and experience. A more experienced team with high skills would know how and when to communicate and what kind of collaboration is needed for the project. Overall, a more experienced and skilled team has more effective communication and better collaboration.

6.8.3 Technical difficulties and complexity were not always related to sustainability

The studied projects show that technical difficulty is not necessarily derived or related solely to the high sustainability and high energy efficiency goals. In case 1's WAF project, the challenge was mainly related to the parallel requirement of historic preservation and an operational goal; the technical complexity was dictated by two factors: the building remained occupied during construction, and the design required many reviews from SHPO because of the building's listing on the National Register of Historic Places.

Regarding the EGWW building in case 2, the technical complexity was mainly caused by the sophisticated layered façade (recladding the entire existing building) and the implementation of a radiant cooling system. The new retrofit façade contributed to energy savings, but the contribution was limited. The radiant cooling system was a mature technology that has been used in the European market for decades. The difficulty was not caused by the system complexity but by the unfamiliarity of this technology in the United States.

For the HMM building in case 3, the target sustainability level was not significant, and the technical complexity was induced by the conflicting needs of modernization and historic preservation. Regarding the FCS building in case 4, the decision to use an underfloor 100% filtered outdoor air system was determined by the project site's adjacency to a cement plant that generated air pollution. Moreover, the curvilinear floor plan created challenges since most floor panels needed to be individually sized and made. Another technical complexity of the structural design was also derived from the curved building form. The curved plan was a design idea that did not relate to energy efficiency nor other types of sustainability performance, and the design team did not provide any explanation of how the plan layout benefited the building's sustainability.

For the BRF building in case 5, the technical complexity of the building façade renovation (recladding) was caused by the requirement of historic preservation and how the precast façade panels were originally manufactured and installed. It was impossible to replace the upper floors' façade panels, so the design team decided to add a new interior façade inside to meet the energy performance goal. This unique solution was created to solve problems related to constructability that many historic buildings encounter.

Regarding the GTML building in case 6, the technical complexities were directly related to the existing building's condition; the imperfections of the initial construction (twisted building and façade) not only prolonged the construction but also accumulated addition cost spending on documenting the existing condition correctly. Besides the existing building's unforeseen condition, maintaining building operations during the construction was also a large technical and logistics challenge.

An examination of the case studies provided solid evidence that the above-mentioned technical complexities were not directly related to any sustainability goals or requirements. **As they occurred in sustainable projects, it is understandable that people struggled to disentangle the project-based complexities from sustainability-based complexities**.

6.8.4 Innovation

The case studies attempted to answer the question derived from the data analysis: Why was innovation less influential in the construction cost? First, innovation defined in projects varies greatly. Certain projects' innovation was related to construction logistics, whereas others related to communicating with existing building tenants. In many cases, they were not specific to sustainability nor to GSA. Innovations can and should be applied to all projects with or without sustainability goals, with the benefit of delivering a better product.

Second, the innovations that occurred in projects were not always driven by the design team – sometimes the design team was resistant to certain innovations due to a lack of knowledge of new technologies. For example, in EGWW, the innovation was driven by the determination of the client rather than by the project designer. The design team maintained a conservative stance since they were not familiar with radiant cooling systems.

Because of the client's (GSA) persistence, the design team eventually hired an additional consultant expert in radiant systems to work with the manufacturer in solving this challenge. The design team's attitude and capacity toward innovation was closely related to their skills and experience, which was the first in-depth finding: team skills and experience were key factors.

6.9 CHAPTER SUMMARY

This chapter presented six case projects for readers to better understand the project characteristics and project team characteristics and their impact on the construction cost. The selected projects represent high, low, and mean construction costs of the projects. The case studies further demonstrate the importance of team skills and experience as dominating factors, together with the procurement method and contractual relations among team members. Furthermore, the case studies provided explanations for the data analysis results (from Chapter 5): why technical complexity and innovation were less influential in the construction cost. The author acknowledges that certain aspects of these project were difficult to replicate for others, however, best practices can be repeated for future projects, such as the selection of a skilled and experienced team and investments in the front-end design. The hope is that these case studies provide the impetus and support for important discussions that will elevate future projects with sustainability goals.

REFERENCES

AIA. (2012). *Edith Green Wendell Wyatt Federal Building Modernization*. http://www.seradesign.com/wp-content/uploads/EGWW-2012-BIM-Awards-Project-Narrative.pdf

AIA Top Ten. (2013). *Federal Center South Building 1202*. AIA. https://www.aiatopten.org/node/204

AIA Top Ten. (2016). *The Edith Green – Wendell Wyatt Federal Building*. AIA. https://www.aiatopten.org/node/494

Apogee Renovation. (2019). *Security, seismic and sustainability Byron Rogers Federal Building and U.S. Courthouse*. https://www.apog.com/static-files/2912fbf4-a353-4102-b59c-bfe78ca17e15

ARCHITECT. (2016). *Westlake Reed Leskosky*. https://www.architectmagazine.com/firms/westlake-reed-leskosky

Architect Magazine. (2010). *Architect Magazine lists SERA Architects as nation's #3 green architecture firm, ZGF as #7 firm overall.* https://chatterbox.typepad.com/portlandarchitecture/2010/05/architect-magazine-lists-sera-architects-as-nations-3-green-architecture-firm-zgf-as-7-firm-overall.html

Bartels, M., & Swanson, M. (2016). *Byron G. Rogers Federal Building: Denver, Colo. From Retro to Retrofit.* https://www.hpbmagazine.org/byron-g-rogers-federal-building-denver-colo/

Chaloeicheep, C. Q., Chatto, C., & Clark, E. (2014a). *14F-Federal-Center-South-Building-1202-Seattle-WA.* High Performing Buildings. https://www.hpbmagazine.org/content/uploads/2020/04/14F-Federal-Center-South-Building-1202-Seattle-WA.pdf

Chaloeicheep, C., Chatto, C., & Clark, E. (2014b). *Shaped to Perform* (High-Performance Building). https://www.hpbmagazine.org/content/uploads/2020/04/14F-Federal-Center-South-Building-1202-Seattle-WA.pdf

Cheng, R. (2015). *Integration at its finest: Success in high-performance building design and project delivery in the federal sector.* University of Minnesota Digital Conservancy. https://hdl.handle.net/11299/201407

Cheng, R. (2018). *Integration at its finest: Success in high-performance building design and project delivery in the federal sector, Volume 2.* Office of Federal High-Performance Green Buildings, GSA.

Cheng, R., Hayter, S., Hotchkiss, E., Pless, S., & Sielcken, J. (2014). *Aspinall Courthouse: GSA's historic preservation and net-zero renovation case study.* https://www.osti.gov/biblio/1163433

Cleveland.com. (2019). *Westlake Reed Leskosky ranked No. 7 out of the top 50 U.S. architecture firms by Architect magazine.* https://www.cleveland.com/architecture/2015/09/westlake_reed_leskosky_ranked_1.html

DCK. (2019, August 23). *DCK worldwide Ranked Among the Top 250 International Contractors and 10th Largest Builder in the Hotel Market Space.* https://www.dckwwgroup.com/about-us

Department of Energy. (n.d.). *Plug & Process Load.* Better Buildings. Retrieved July 7, 2022, from https://betterbuildingssolutioncenter.energy.gov/alliance/technology-solution/plug-process-loads

DLG Group. (n.d.). *Howard M. Metzenbaum U.S. Courthouse.* https://www.dlrgroup.com/idea/awa-cleveland-adventure/

DOE. (2009). *2009 American Recovery and Reinvestment Act.* https://www.energy.gov/oe/information-center/recovery-act

Edzarenski. (2021). *Construction Analytics Economics behind the Headlines.* Construction Inflation May 2021. https://edzarenski.com/2021/05/21/construction-inflation-report-may-2021/

EIA. (2022). *Annual Energy Outlook 2020.* U.S. Energy Information Administration. https://www.eia.gov/outlooks/aeo/

ENR. (2019). *ENR 2019 Top 100 Green Building Contractors.* https://www.enr.com/toplists/2019-Top-100-Green-Building-Contractors

EPA. (2007). *Summary of the Energy Independence and Security Act.* https://www.epa.gov/laws-regulations/summary-energy-independence-and-security-act

General Services Administration. (2002). *Treasury, Postal Service and General Government Appropriations for Fiscal Year 2002.* General Services Administration.

General Services Administration. (2014). *Wayne N. Aspinall Federal Building and US Courthouse.* https://www.gsa.gov/about-us/regions/welcome-to-the-rocky-mountain-region-8/buildings-and-facilities/colorado/wayne-n-aspinall-federal-building-and-us-courthouse

Gerardi, J. (2021). *Commercial construction costs per square foot.* https://proest.com/construction/cost-estimates/commercial-costs-per-square-foot/

GSA. (n.d.). *Byron G. Rogers Federal Building and U.S. Courthouse, Denver, CO.* https://www.gsa.gov/historic-buildings/byron-g-rogers-federal-building-and-us-courthouse-denver-co

Howard M. Metzenbaum U.S. Courthouse. (n.d.). Retrieved September 10, 2021, from https://cbe.berkeley.edu/wp-content/uploads/2018/07/metzenbaum-submittal.pdf

Lobato, C., Sheppy, M., Brackney, L., Pless, S., & Torcellini, P. (2012). *Selecting a Control Strategy for Plug and Process Loads* (NREL/TP-5500–51708, 1051936; p. NREL/TP-5500–51708, 1051936). https://doi.org/10.2172/1051936

Mørck, O., Sánchez, M., Lohse, R., & Riel, M. (2017). *Deep Energy Retrofit – Case Studies. Business and Technical Concepts for Deep Energy Retrofit of Public Buildings Energy in Buildings and Communities Programme Annex 61, Subtask A.* International Energy Agency. https://iea-ebc.org/Data/publications/EBC_Annex%2061_Subtask_A_Case_Studies.pdf

Park, J., & Kwak, Y. H. (2017). Design-Bid-Build (DBB) vs. Design-Build (DB) in the U.S. public transportation projects: The choice and consequences. *International Journal of Project Management*, *35*(3), 280–295. https://doi.org/10.1016/j.ijproman.2016.10.013

Rimmer, J. (2013). *Architects Warm to Chilled Ceilings.* https://continuingeducation.bnpmedia.com/courses/price-industries/architects-warm-to-chilled-ceilings/3/

Smart Remodeling LLC. (2022). How Do You Calculate Commercial Renovation Costs. *Smart Remodeling.* https://www.smartremodelingllc.com/blog/how-do-you-calculate-commercial-renovation-costs

Tupper, K., Hammer, N., Osbaugh, R., & Swanson, M. (2012). Right Steps for Retrofits: Byron G. Rogers Federal Office building case study. *ASHRAE Transactions*, *118*, 11–18.

Walter P Moore. (2014). *Mickey Leland Federal Building Renovations.* https://www.walterpmoore.com/projects/mickey-leland-federal-building-renovations

WBDG. (2017). *Federal Center South Building 1202* [Case study]. Whole Building Design Guide. https://www.wbdg.org/additional-resources/case-studies/federal-center-south-building-1202

7   The path forward

ABSTRACT

Built upon the analysis results from the empirical data (Chapter 5) and in-depth case studies (Chapter 6), this final chapter addresses three areas. First, it uses empirical data to debunk the perception of sustainable building being more expensive than conventional building. Second, it provides insight into the root causes of this misperception. Lastly, it outlines a path forward to affordable green building using examples.

7.1 DEBUNKING THE "EXPENSIVE" PERCEPTION

The level of sustainability (LOS), or the degree of greenness, has been perceived as a main driver for the higher construction cost of sustainable building. However, the data analysis results (refer to Chapter 5) revealed that despite the general consensus, the relation between LOS and cost is inconclusive. When comparing the actual construction cost of the GSA LEED buildings to conventional buildings, the average new LEED building's construction cost was lower than the average cost of conventional (mid-rise, non-LEED) office buildings: $4,722/m^2 vs. $6,244/ft^2 (Gerardi, 2021). The average LEED renovation project construction cost was also lower than that of conventional buildings: $2266/m^2 vs. $5000/m^2. These findings indeed differ from the common perception and previous studies based on survey data (refer to Chapter 4). Moreover, other commonly perceived construction cost drivers, such as technical complexity and project type, also did not exhibit a causal relation to the construction of GSA LEED buildings (refer to Chapter 5).

While initially difficult to understand why the commonly perceived factors were noninfluential based on the empirical data, the case studies provided insight. For example, it was found that technical complexity and technical difficulty were mainly induced by two resources: the project teams' unfamiliarity with certain technology (not necessarily new) and the restrictions and constraints related to other project requirements, such as historic preservation requirements. These types of technical complexity and/or difficulty can be well mediated by acquiring experienced and skilled team members who are familiar with the systems and regulations. This begs the question of whether the "expensiveness" of sustainable building is "manmade" and thus avoidable or, at a minimum, can be mitigated or controlled. The key to the answer is the often-neglected project team characteristics, especially in the early design phases.

The effect of the project team characteristics can be interpreted as follows: the **higher the level of skill** sets of a project team, the **more experience the project team** has with sustainable buildings, and the **more effective communication** the team has, the **lower the construction cost** related to the high LOS the project achieves. Among the studied team characteristics, skills and experience were dominating factors, which aligns with the common perception. However, the magnitude of the impact the design cost (paid to project team) had on the final construction cost is almost equal to that of hard costs. In this context, a hard cost is defined as a type of direct cost that can be traced back to the physical components or activities of the construction project. Hard costs include the costs for labor, materials, and equipment and the overhead costs from the contractor and developer.

Since the design cost is less than 10% of the total construction cost, it may be difficult to comprehend how the design cost carries the same influence as the remaining 90%. For clarity, the design cost impact is referring to the influence on fluctuations (changes) in the total construction cost, rather than the absolute monetary value. To better understand, a pancake metaphor can be used: pancakes are normally made of two portions of flour, two portions of milk, one portion of eggs, and half a portion of butter. While half a portion of butter represents a small amount of all the ingredients combined, it is the ingredient with the most cost and taste fluctuations, influenced by many unforeseen circumstances. Consequently, butter can drive most of the cost changes for making pancakes and alter the taste of the pancakes. This metaphor relates to the role of the design cost: even if it is a small portion, it will drive most cost variations in the sustainable building construction and increase (or decrease) the LOS of projects.

However, until now, the project team's influence during the early design stage on the total construction cost has not been given enough attention in comparison to the focus on the project team's performance during the construction phase (e.g., construction productivity). Put simply, the common belief is that the construction team (contractors), rather than the project design team, is more important for construction cost control. More specifically, due to its small portion, the design team's qualifications have not been regarded widely as a key factor in driving the cost variance of sustainable building. The findings from the empirical data and case studies of the GSA LEED projects support the importance of the design team. These findings have a theoretical foundation based on the concept of the effort curve, which was proposed decades ago (refer to Chapter 1): *The most important decisions related to project risks and management are often decided in the early design phases, and these risks can lead to a cost overrun if not managed appropriately*. The effort curve has been understood with conventional anecdotal evidence, or simply intuitively. The empirical evidence provided in this book, using actual construction costs and in-depth case studies, may be a first with respect to sustainable building in the United States. Several general recommendations can be made about controlling the sustainable building construction cost (SBCC):

> Overall, the collective findings from the data analysis and case studies strongly support the notion that **acquiring the right project team** carries more weight than solving technical complexity in controlling the total construction cost of sustainable buildings. In addition, the added construction costs of sustainable buildings can be avoided **without compromising** the level of sustainability achieved.

7.2 ORIGINS OF THE MISPERCEPTION

7.2.1 Lack of empirical data

A lack of data is one important cause for why the misperception has lingered for so long without being adjusted. The current common perception of a high SBCC is largely derived from the survey results, and even from the meta-analysis conducted by the author, using the academic literature. Among the included 31 studies, only 21 % of the studies were able to obtain the actual construction cost data and documents from the project team, while the remaining studies relied on survey or questionnaire responses from project team members (architects, interior designers, and engineers) and clients or developers. The absence of studies based on actual building construction costs may indirectly contribute to the perception of green building costing more.

While the empirical findings from this book may be the first in the United States, they align with another recent independent study in Europe that examined 37 DGNB certified projects. DGNB, or the German Sustainable Building Council, is a non-profit organization founded in 2007, based in Stuttgart. The 37 buildings included eight multistory residential buildings, 21 office buildings and eight terrace houses. All projects are located in Denmark and were built between 2012 and 2019 (BUUS Consult, 2020). The total construction costs of the projects range from €800/m^2 ($955/m^2) to €3,200/m^2 ($3,833/m^2), which are comparable to the GSA projects included in this book (Kehl, n.d.). The DGNB study results indicate that sustainable buildings are not necessarily more expensive. Neither higher DGNB award levels nor lower environmental impact levels were necessarily associated with a higher construction cost. This study is one of few studies showing no correlation between the SBCC and LOS.

As mentioned in Chapter 1, an overarching research aim of this book is to propose a framework to understand the underlying drivers and factors of the SBCC. The newfound framework (refer to Chapter 5's structural equation modeling) can be used to analyze any collected data on actual SBCCs beyond LEED and GSA buildings. The more empirical data on the SBCC that can be collected and analyzed, the better the public can understand the true cost of sustainable building.

7.2.2 Sustainability and unconventional luxury

Why do the public and many professionals still perceive green building as expensive? Why is the thought of sustainable building being expensive acceptable? The theory behind luxury brands can provide insight to these questions. As a brand, sustainable building has been associated with being environmentally friendly and socially responsible. In this case, sustainable building is perceived as an "unconventional luxury" product. The idea of unconventional luxury is defined as a concept representing something *more* than just product quality (Han et al., 2010); rather, it is to a large extent a social construct with perceived benefits beyond product quality, and it changes with societal development (Kapferer & Bastien, 2009). Sustainable buildings, especially LEED certified buildings, fit perfectly into this social construct.

Consumers, or the public, associate sustainable buildings with long-term and societal benefits beyond the individual building footprint. Unlike traditional luxury goods with product-focused views, unconventional luxury products focus on how luxury is experienced. Thomsen et al. (2020) provided a great explanation of the difference between conventional luxury and unconventional luxury: in traditional conceptualizations, luxury is often presented as being a limited resource, while in unconventional conceptualizations the luxury is presented as being only restricted by the perceptual abilities of the consumer. Consequently, traditional luxury products often have a higher price tag derived from limited resource materials (e.g., rare metals), while the high price of unconventional luxury products is associated with consumers' perception (e.g., organic food is healthier than regular produce).

How does unconventional luxury drive up the perceived cost of sustainable buildings? General consumers do not have knowledge of the difference between the price and actual cost of sustainable buildings since the cost data are often hidden among all other costs. Therefore, public consumers, the actual building users, have assumed that a linear relation exists between the price and cost. In elitist consumption, high prices offer symbolic measures of the consumer's own value, achievement, and status (Berry, 1994; Allsopp, 2005). Products and prices can improve social standing, but this benefit disappears as more people gain access to those luxury brands. The benefit can be maintained through raising the price or keeping the perception of the product being excellent and expensive. The luxury industry's desire is to present itself as a manufacturer of excellence, which is often time associated with distinctive aesthetics and unique artisanal quality because the impact of internal quality takes much longer to build (Kapferer & Valette-Florence, 2021). In sustainable buildings, this quality and excellence that are perceived and reflected on the building's appearance only partially contribute to its sustainability. A quick internet search using the keyword "sustainable buildings" results in images of luxury office buildings with green

Figure 7.1

Google image search using the keyword "sustainable buildings"

roofs, green walls, or unique designs. As shown in Figure 7.1, an ordinary-looking office building or single-family building does not appear in the first 35 images.

It is imperative to point out that no sustainable building certification criteria requires the building to be wrapped with greenery. Furthermore, the energy-saving benefits of a green roof (compared to a conventional roof) vary tremendously, from 0.7% (Sailor et al., 2012; EPA, 2015) to 65.3% (Bevilacqua, 2021). A simple Google search demonstrated the public image of sustainable buildings, which is reflective of the public perception of "sustainability": being somewhat **unique** and **visual** and being a **luxury** item. **What is worrisome is that none of these characteristics actually define sustainability**.

7.2.3 Sustainability through a visual lens

With the connection between luxury and sustainability defined, the following questions focus on who is promoting this notion and how it can be corrected. The simple answer is everyone is promoting the notion, and it must be first stopped from within the professional community. For example, an article published on Mansion Global[1] featuring an eco-focused designer was titled "Sustainability and Luxury Are Made for Each Other, Says Eco-Focused Designer" (Kaminer, 2022). According to the website profile, this eco-designer is an expert in circular design and an advocate for advancing sustainable and healthy materials. She defined luxury "as being ethical, while designing or surrounding oneself with forward-thinking products and practices, aesthetically pleasing with circularity, health and wellness in mind" and she describes her favorite sustainable luxury materials as "Irish and Australian wools" and "mushroom-based" leather (Kaminer, 2022). It is foreseeable that using Irish and Australian wools for a small residential project located in the Midwestern United States can have significant implications on the added costs. More importantly, considering the large eco-footprint of Australian-sourced materials and products being transported and used in the United States questions its sustainability from the perspective of a whole life cycle impact. This example is just one of many episodes leading to the deep misperceptions around sustainable building in the public eye, with embodied energy such as shipping being the least visible, though a significant contributor.

It is understandable that designers focus on the visual and aesthetical appearance of sustainability since "designers are trained to be a visual communicator and believer." In fact, most people are visual consumers in this social media-dominating era. A notion, a concept, or a movement cannot be spread if it is not aesthetically appealing, pleasing, or intriguing. Consequently, when people try to promote sustainability, flashy or eye-catching photos are often used to catch public attention. As a result, a website featuring a net-zero energy building or sustainable house appears expensive, with trendy zero-waste accessories and minimal wood furniture, among others. Society often connects sustainability with a certain appearance, and these ideas can be perpetuated by alleged experts in the professional community.

7.3 THREE PRINCIPLES OF A PATH TOWARD AFFORDABLE SUSTAINABLE BUILDING

To combat the misperception of sustainable building being expensive, a luxury item, and something special or unique, three components should be promoted for a path toward understanding sustainability: sustainability is **universal**, a **baseline,** and **non-visual**. To promote these principles and make sustainable building affordable, efforts can be made from three areas to engage all stakeholders: legislation, practice, and finance. In the following sections, the author will outline a path forward.

7.3.1 Legislation: sustainability as mandatory (universal)

To make sustainable building a universal practice, a top-down approach has proven to be more effective than a bottom-up approach. The results from the GSA LEED buildings included in this book provide solid evidence that the top-down mandate from GSA is highly effective in terms of turning the sustainability goal into reality. In Chapter 6, the case studies demonstrate that all building types (new construction and renovation), all locations (from the West Coast to the East Coast), and all building sizes (3 stories to 22 stories) can meet the sustainability goal within a cost that is comparable to (or even lower than) their conventional counterparts.

Next, two scenarios are established to investigate how rapidly the building stock can be transformed into net-zero energy buildings in three countries: Germany, the United States, and China. These countries were chosen because they represent differences in existing building stock conditions, regulation approaches, existing building energy code requirements, and existing energy user behavior. Together, the three countries represent a useful cross-section of existing stock conditions and future global trends, presenting a comprehensive overview of world conditions (Hu & Qiu, 2018).

Scenario one, the reference scenario, serves as the baseline and assumes that all building codes and policies initiated up to 2018 in the three countries will continue to impact building energy demand. No additional policies and code changes are to follow 2022, although autonomous technological improvements are expected to occur through 2050, with a typical building life span of 50 years. The baseline year 2018 was chosen because the following building codes of that year are more comparable among the three countries: ASHRAE 2016 (US), the Energy Saving Ordinance (EnEV) 2016 (Germany), and the Chinese 2018 Building Code.

Figure 7.2 presents the reference scenario (baseline) based on current building codes and policies related to building energy efficiency design without future changes. The United States has the highest annual energy consumption per building floor, with energy usage intensity (EUI) at 49 kBTU/ft^2/yr (156 KWh/m^2/yr); China is second with 45 kBTU/ft^2/yr (143 KWh/m^2/yr), and Germany is the lowest with 29 kBTU/ft^2/yr (93 KWh/m^2/yr). Current building

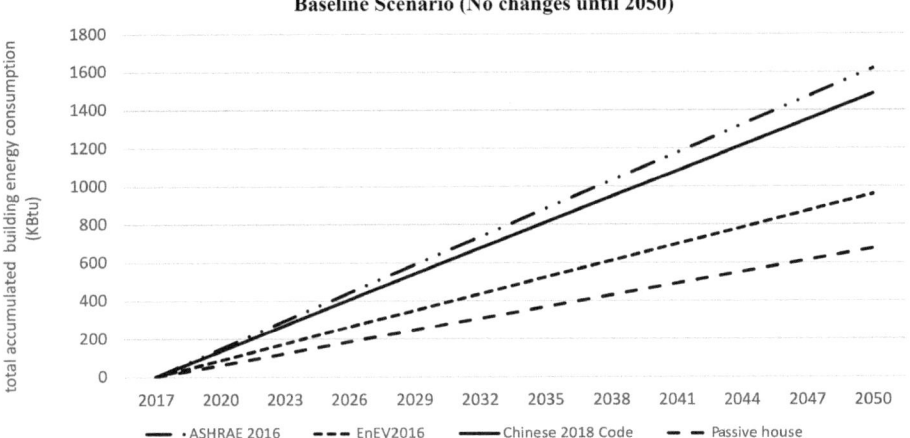

Figure 7.2
Reference scenario

energy codes in China are much less stringent than in the United States and Germany; however, building energy intensities (EUIs) in China are lower than in the United States due to the operational schedule, occupancy behavior, and cultural context. The operational schedules in China and Germany are much shorter than in the United States. For example, many office buildings in the two countries will turn off air-conditioning systems completely after normal working hours, while US office buildings only lower the settings to maintain a certain minimum level of air-conditioning. Therefore, even with a lower build-ing energy efficiency requirement, China still has a lower EUI than the United States. As illustrated in Figure 7.2, if no changes are made to current building energy standards regarding cumulative energy use, until 2050, the United States could consume 69% more energy than Germany, 9% more than China, and 140% more than the current Passive House standard. Consequently, the United States will have the highest annual energy consumption per building floor (Hu & Qiu, 2018).

The second scenario, the path forward scenario, represents a vision of transforming to a carbon-neutral, net-zero energy economy by 2050. Currently, Germany and China use top-down approaches—one centralized policy govern-ing most building energy code developments and upgrades. The EU has set itself a long-term goal of reducing greenhouse gas (GHG) emissions by 80–95% by 2050, compared to 1990 levels (European Environment Agency, 2015). The German government has implemented policies both for new and existing build-ings to better conserve energy by 2020 and 2050. Germany has the goal of reducing GHG emissions by 95% by 2050 compared to the 1990 baseline, fol-lowing the EU directives. The building codes have been strengthened five times over the past 35 years (1990–2022), and an energy demand reduction for space heating and domestic hot water has been achieved, from 300 to 65 kWh/m²

(Schimschar et al., 2011). The transformative scenario in China was set to convert the country's building sector into a self-sustaining economy with increased energy efficiency through 2050. On December 10, 2016, the National Energy Administration adopted its thirteenth Five-Year Plan (2016–2020) and established a series of energy targets. The key objectives include an increase of 15% and 20% in non-fossil energy in total energy consumption by 2020 and 2030, respectively, and plans to reduce China's emission intensity by 2020, at a higher rate than other major economies. For the path forward scenario, the rates from the thirteenth Five-Year Plan were used: a 30% increase every six years, with the goal to reach net zero by 2030 (Lo, 2014).

Lastly, in the United States, different jurisdictions have the right to adopt and develop their own codes and regulations. Currently, there are no nationwide building code requirements related to a net-zero goal, and certain local jurisdictions have their own targets. In this comparison, the most progressive policy has been adopted for the US transformative scenario. California Title 24 of the Energy Efficiency Standards for Residential and Non-residential Buildings set the following net-zero energy building goals: all new residential construction will be zero net energy (ZNE) by 2020, all new commercial construction will be ZNE by 2030, 50% of commercial buildings will be retrofit to ZNE by 2030, and 50% of new major renovations of state buildings will be ZNE by 2025 (California Public Utilities Commission).

Figure 7.3 indicates a transformative scenario with major policy influence. Germany has had the lowest overall energy consumption in 60 years because

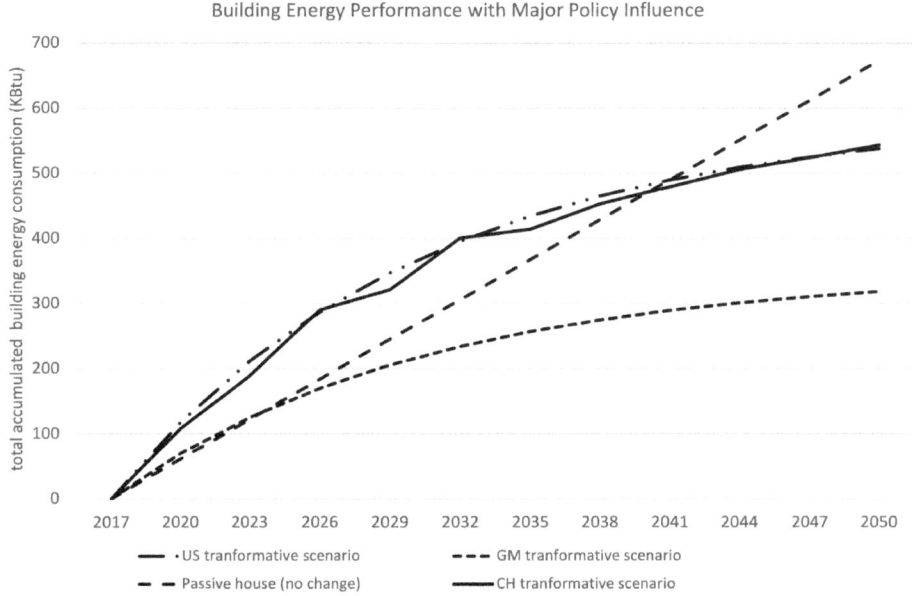

Figure 7.3
Transformative scenario with major policy influence

of two factors: the early adoption of a building code and an efficient operation schedule, and the mandatory requirement of reducing energy use by 20% by 2020 and 50% by 2050. Overall consumption from the German building industry could become relatively flat after 2035. China has the second lowest overall energy consumption, consuming only 40% of energy compared to the baseline, and it has the highest energy efficiency increase among the other countries. Compared to the reference scenario, regarding energy savings, China could save 60.5%, Germany 65%, and the United States 17%. By approximately 2038, the Chinese building energy efficiency standard could bypass the existing Passive House standard, and by 2050, China's building energy efficiency would become more comparable to Germany's. These significant changes would be the result of continuously improved energy efficiency under China's five-year strategic plans and 2020 and 2030 goals. The most significant turning points would be around 2025 and 2035, where the full impact of the 2020 and 2030 goals could be fully reflected. In the United States, the building efficiency standard could bypass the existing Passive House standard by 2040, and by 2050, the efficiency would be almost the same as that of China (Hu & Qiu, 2018).

The analysis and modeling results demonstrate that implementing top-down policies to encourage building efficiency improvement is essential to helping different countries achieve ultimate sustainability goals in terms of energy performance. For all three countries, the effects are significant. The analysis results clearly indicate that around 2040, the United States' energy efficiency standard will be able to outperform the Passive House standard if all states follow California's standard under the transformative scenario. Conversely, without the mandated policy, as illustrated in Figure 7.2, the sustainability goal is difficult to reach.

7.3.2 Professional requirement: sustainability as a design baseline

After several decades of development, energy-efficient practices and sustainable building design have become a choice – a sectarian one (Hu, 2019). Sustainable living has become a style and the choice of certain groups. The GSA LEED case studies demonstrated the importance of skills and experience in the design team, as not all teams have sustainable design skills. However, the principles of sustainable design are largely based on the laws of physics and energy flow, and these principles have not changed since their establishment centuries ago. These sustainable design principles have been intuitively applied in many traditions worldwide and have been used by designers and craftsmen to adapt to different local climatic conditions. **They are not new; instead, those principles are ancient, historical, and universal**; for example, orienting a building and designing its fenestration and external shading devices to correctly maximize the heat intake from the sun during winter while minimizing solar heat exposure during summer.

So why has sustainable design suddenly become a new and specialized design approach known to a small group of people? Can sustainable design principles become universal guidelines again? Changing the status of sustainable design as a specific skill set owned by a certain group of experts requires an overhaul of the education system to integrate **sustainability as a fundamental design principle,** *rather than as an expertise or skill that can be acquired later and individually. If sustainable design principles were to become part of the basic design vocabulary, then the next generation of architects, engineers, and contractors would be able to employ these principles in every project. It is imperative to mention that most sustainable design principles are passive, low cost, and even no-cost. For example, orienting a building to be north-south facing and providing enough operable windows for natural ventilation do not add any additional costs.*

The current perception and approach to achieving the sustainability goal heavily depends on advanced and expensive technology, which has already created huge financial and technological barriers for many less-resourceful projects and building owners. Furthermore, some designers and developers have been intentionally misinterpreting sustainability with luxury. What is needed is a knowledge-based sustainable design education approach.

If education is the foundation to ensure sustainable design becomes the baseline for all projects, then ethics training and disciplinary requirements are the lateral system to ensure the sustainable design approach can withstand any external resistance. In the current professional community, practitioners rarely proactively push forward a practical and sustainable design approach if there is no interest from owners or incentives from legislation.

Regarding disciplinary requirements, the design of a building's environmental control system (heating and cooling) can be used as an example. Mechanical heating and cooling systems in buildings use energy no matter how efficient they are. The greater the dependency upon them, the more energy is consumed in the building. The architect has the first opportunity to provide proper thermal conditions by using the building itself (geometry, orientation, building envelope). Any remaining needs can be satisfied by the mechanical systems. Architects must accept the responsibility that belonged to them traditionally to minimize the building's dependency on a heating and cooling system. Additionally, architects are called upon to return to the roots of a climate-sensitive design approach. Energy-conscious design requires that the total energy use of the building be considered from its inception.

The fundamentally misleading concept of contemporary sustainable building is that its building type differs from others. The label of sustainability is used to differentiate buildings as consumer products, and this division is profit- and market-driven. However, when the origin of sustainability is traced back to its ecological roots, sustainable building is understood as not one building type or one design approach, but as a guiding **design principle for all buildings** and **professional ethics for all practitioners**. Just as all mechanical designs must follow the laws of thermodynamics, all building designs should treat the sustainability concept as the core consideration, and not as an add-on item" (Hu, 2019). Only through design education and ethics training for professionals can sustainability become a baseline practice rather than a voluntary aspiration to be achieved by certain groups.

7.3.3 Investors and developers: sustainability as risk management

As mentioned in Chapter 1, the size and scale of "brown discounts" in the market have been rising rapidly in recent years. Brown discounts refer to buildings that fail to achieve environmental standards and are less energy-smart, making them seen less valuable. Increased market value and resale value have been the main drivers of the green building market. It is encouraging to see that traditional concepts of real estate, such as being in a prime location, are gradually shifting toward environmental responsibility and the promotion of health and well-being. In 2020, global investment in sustainable building increased by 11.4%, around $184 billion, dominated by European countries. However, in the United States, sustainable building and energy efficiency investments only increased by 3% in 2020, which indicates the United States is not a major contributor to global green construction activities (United Nations Environment Programme, 2021).

Regardless of the positive changes, sustainable building is still considered a minority (Ulbrich, 2022). Waiting for building regulations or national policies to catch up is not adequate for urgent needs. A recent survey report "Decarbonizing the Built Environment: ambitions, commitments and actions" published by JLL (NYSE:JLL) in 2021, surveying 647 building investors and residents globally (US, China, UK, France, Germany, Japan, and Australia) indicates that leading real estate developers and investors are outpacing policy-makers in their commitment to sustainable buildings. They take their own initiative and make changes primarily because they firmly believe it is the "right thing to do" for their stakeholders – which include customers, employees, and communities – rather than being regulated to do so. Furthermore, they demand to see national and city governments take bolder steps to support them through regulatory incentives (JLL, 2021). *Another important finding reported is that climate risk is now being accepted and viewed as a financial risk*. Researchers found that 83% of real estate occupiers and 78% of investors believe that climate risk poses a financial risk. In January 2020, CEO Larry Fink of BlackRock – an American multinational investment management firm and the world's largest asset manager – stated in his annual letter, "The evidence on climate risk is compelling investors to reassess core assumptions about modern finance.... Climate risk is investment risk... climate change has become a defining factor in companies' long-term prospects" (Fink, 2020). His letter stimulates a growing debate over whether building green should be treated as a risk management technique over the outdated mindset of treating it as a profitable asset. If investors and developers adopt building green as a fundamental practice for risk management, then sustainable building can quickly become the baseline for individual decisions. However, developers cannot do this alone; as Fink pointed out, "Capitalism has the power to shape society and act as a powerful catalyst for change. But businesses can't do this alone, and they cannot be the climate police" (Mufson & MacMillan, 2022). As a market-driven approach cannot replace the function of governmental regulation, policy-makers, the professional community, and developers/investors need to work together.

7.4 AN EXAMPLE OF A PATH FORWARD

Even with the three key principles outlined for moving toward affordable sustainable building – a universal mandate, professional requirements, and business management – it may still be difficult to imagine the path forward. In this section, the author provides an example of the private and public sectors working together to establish a goal, allocate funding, and share the risk. The example is Singapore's Green Building Masterplan and green finance provided by the government.

Green finance refers to the financial products and services that engage capital markets to create financial support for activities with positive environmental outcomes. The United Nations Environment Programme defined green financing as intending to "increase the level of financial flows from public, private and not-for-profit sectors to sustainable development priorities" (United Nations Environment Programme, n.d.). Green finance in buildings and construction includes green loans and green bonds for developers and building owners. Singapore's Green Building Masterplan was first introduced in 2006 by the Building and Construction Authority (BCA), a governmental agency, to encourage, enable, and engage industry stakeholders to adopt new green building standards. The initial target was new buildings, with the main goal being to encourage developers to adopt sustainability as part of the standard practice. In a later revision, BCA expanded the scope to include the existing building stock, and stakeholders were also expanded to include building occupants with an aim of changing their energy consumption behavior.

The Green Building Masterplan has three key targets for 2030. The first target is for 80% of buildings (by gross floor area) to be green (certified) by 2030, including new construction and existing building retrofits. To encourage existing building energy retrofits, BCA plans to launch an additional $63 million (US dollars) cash incentive scheme in the second quarter of 2022 to help building owners lower the upfront construction cost of existing building energy retrofits (Singapore Building and Construction Authority, n.d.).

The second target is for 80% of new buildings (by gross floor area) to become Super Low Energy (SLE) buildings from 2030. (SLE is defined as the building with 60% energy saving above 2005 building codes which is being used as the anchor reference for Singapore Green Mark energy saving) (Singapore

Building and Construction Authority, 2022) To help achieve this goal, a bonus gross floor area (GFA) was put in place to incentivize developers and building owners to develop SLE buildings. This incentive was launched in November 2021, and buildings that achieve Green Mark Platinum SLE with Maintainability Badge can enjoy up to 3% additional GFA allowed beyond the Master Plan Gross Plot Ratio (Singapore Building and Construction Authority, n.d.).

The third target is to achieve an 80% improvement in energy efficiency for best-in-class green buildings by 2030. The current (2021) best-in-class green buildings can achieve more than a 65% improvement over a 2005 baseline. To move toward this higher bar, BCA will provide an additional $45 million (USD) for this target.

From the above description, readers can understand the Singapore government has aligned financial support with key policies to work alongside all stakeholders in the green building market. To the author's knowledge, this approach is not currently seen in any jurisdiction in the United States. Besides green finance, the Singaporean government also has the Building Retrofit Energy Efficiency Financing (BREEF) program, which offers financing for upfront costs of retrofits through an energy performance contract. For this program, the BCA is responsible for 40% of the risk of any loan default with participating financial institutions under the condition that the credit facility leads to the existing building achieving the Green Mark Certification standard, which is maintained through loan tenure (United Nations Environment Programme, 2021).

While establishing goals and providing financial support are important, it is also necessary to hold developers, building owners, and building occupants accountable to ensure buildings perform and to avoid greenwashing. Disclosing and reporting building energy performance is an effective way. The Singaporean government, through BCA, has required that all building owners submit mandatory annual energy performance data to the Building Energy Submission System for use by BCA in developing energy performance benchmarks. The energy performance data are released each year through Singapore's open data platform (BCA, n.d.), and the data are publicly accessible.

Singapore's combined Green Building Masterplan and green finance is a practical path forward that can be adjusted and implemented in other countries. Other than Singapore, European countries are also making an effort to combine public and private forces to invest in green buildings under national policy and mandates to reduce carbon emissions. For example, Germany has doubled its energy efficiency expenditure in buildings. KfW Development Bank, one of world's leading promotional banks, working on behalf of the Federal Republic of Germany and federal states, has programs for energy-efficient construction and refurbishment that ramp up to €30 billion (United Nations Environment Programme, 2021). In 2020, the French government announced nearly €7 billion for energy efficiency improvements in private homes, office buildings, and public buildings, such as schools and town halls. Furthermore, the United Kingdom's investment in energy efficiency has increased by 18% in 2020 under the joint program between the government and private banks (United Nations Environment Programme, 2021). The examples from these countries demonstrate how the joint force between public policy-makers and private investors is the path toward making sustainable buildings affordable for everyone, regardless of a country's size or political structure.

7.5 LESSONS LEARNED

The intent of this final chapter has been to debunk the common perception of sustainable building being costly, followed by an explanation of the misperception's three root causes: **a lack of empirical data and transparency, sustainability treated as a luxury product, and sustainability viewed from a visual lens**. Moreover, certain professionals working on sustainable buildings also directly or indirectly misguided the public and promoted the notion of sustainable building being expensive. Empirical data collected from the GSA LEED buildings demonstrates that sustainable buildings are not necessarily more expensive than conventional buildings, and they can also be cheaper if executed properly. In addition, the data analysis shows a skilled project team plays an important role in mitigating and avoiding unnecessary construction costs. Therefore, training professionals to be equipped with the necessary skills and mindset is one important component toward affordable sustainable building practices. The other important components are legislative requirements from policy-makers and using sustainability as risk management investment from investors. The path forward is not easy, but several countries are advancing in the right direction, with Singapore as a seminal example.

> The causes identified in the book for the misperception of sustainable design being more expensive do not sufficiently answer the question of how to make sustainable building more accessible to everyone. However, it provides a foundation by correcting several misunderstandings fuelled by a lack of empirical data. While it is impossible to change public perception with one data set, this book's attempt is to shift the perception of sustainable building as a luxury brand to its being recognized as a baseline by all involved in the process. The ultimate hope of this book is for readers to clearly understand that sustainability is a set of underlying design principles that value the right mindset over special technologies or materials.

NOTE

1   A digital destination for content about the global real estate market.

REFERENCES

Allsopp, J. (2005). Additional practice papers: Premium pricing: Understanding the value of premium. *Journal of Revenue and Pricing Management, 4*(2), 185–194. https://doi.org/10.1057/palgrave.rpm.5170138

BCA. (n.d.). *Benchmarking Reports and Data*. Bess (Building Energy Submission System. Retrieved June 17, 2022, from https://www.bca.gov.sg/BESS/BenchmarkingReport/BenchmarkingReport.aspx

Berry, S. T. (1994). Estimating discrete-choice models of product differentiation. *The RAND Journal of Economics, 25*(2), 242. https://doi.org/10.2307/2555829

Bevilacqua, P. (2021). The effectiveness of green roofs in reducing building energy consumptions across different climates. A summary of literature results. *Renewable and Sustainable Energy Reviews, 151*, 111523. https://doi.org/10.1016/j.rser.2021.111523

BUUS Consult. (2020). *Is it expensive to build sustainable? – Report by BUUS consult on the relation between building costs and sustainability*. https://dk-gbc.dk/publikation/is-it-expensive-to-build-sustainable

EPA. (2015). *Using Green Roofs to Reduce Heat Islands*. U.S. Environmental Protection Agency. https://www.epa.gov/heatislands/using-green-roofs-reduce-heat-islands#:~:text=In%20addition%2C%20green%20roofs%20can,foot%20of%20the%20roof's%20surface

European Environment Agency. (2015). *Mitigating climate change*. https://www.eea.europa.eu/soer/2015/europe/mitigating-climate-change#:~:text=The%20EU%20aims%20to%20decarbonise,to%201990%20levels%20by%202030

Fink, L. (2020). *A Fundamental Reshaping of Finance—Larry Fink's 2020 Letter to CEOs*. https://www.blackrock.com/us/individual/larry-fink-ceo-letter

Gerardi, J. (2021, June). *Commercial construction costs per square foot*. https://proest.com/construction/cost-estimates/commercial-costs-per-square-foot/

Han, Y. J., Nunes, J. C., & Drèze, X. (2010). Signaling status with luxury goods: The role of brand prominence. *Journal of Marketing, 74*(4), 15–30. https://doi.org/10.1509/jmkg.74.4.015

Hu, M. (2019). *Net zero energy building: Predicted and unintended consequences*. Routledge. https://doi.org/10.4324/9781351256520

Hu, M., & Qiu, Y. (2018). A comparison of building energy codes and policies in the USA, Germany, and China: Progress toward the net-zero building goal in three countries. *Clean Technologies and Environmental Policy*, 1–15.

JLL. (2021). *Decarbonizing the Built Environment*. https://www.us.jll.com/en/trends-and-insights/research/decarbonizing-the-built-environment

Kaminer, M. (2022, May 30). *Sustainability and Luxury Are Made for Each Other, Says Eco-Focused Designer*. https://www.mansionglobal.com/articles/sustainability-and-luxury-are-made-for-each-other-says-eco-focused-designer-01653906068?mod=

Kapferer, J.-N., & Bastien, V. (2009). The specificity of luxury management: Turning marketing upside down. *Journal of Brand Management, 16*(5–6), 311–322. https://doi.org/10.1057/bm.2008.51

Kapferer, J.-N., & Valette-Florence, P. (2021). Which consumers believe luxury must be expensive and why? A cross-cultural comparison of motivations. *Journal of Business Research, 132*, 301–313. https://doi.org/10.1016/j.jbusres.2021.04.003

Kehl, L. (n.d.). The myth of increased costs in sustainable building. *DGNB*. Retrieved June 17, 2022, from https://blog.dgnb.de/en/myth-of-increased-costs-in-sustainable-building/

Lo, K. (2014). A critical review of China's rapidly developing renewable energy and energy efficiency policies. *Renewable and Sustainable Energy Reviews, 29*, 508–516. https://doi.org/10.1016/j.rser.2013.09.006

Mufson, S., & MacMillan, D. (2022, January 18). BlackRock's Larry Fink tells fellow CEOs that businesses are not 'climate police.' *The Washington Post.* https://www.washingtonpost.com/climate-environment/2022/01/18/blackrock-larry-fink-letter-climate/

Sailor, D. J., Elley, T. B., & Gibson, M. (2012). Exploring the building energy impacts of green roof design decisions – a modeling study of buildings in four distinct climates. *Journal of Building Physics, 35*(4), 372–391. https://doi.org/10.1177/1744259111420076

Schimschar, S., Blok, K., Boermans, T., & Hermelink, A. (2011). Germany's path towards nearly zero-energy buildings—Enabling the greenhouse gas mitigation potential in the building stock. *Energy Policy, 39*(6), 3346–3360. https://doi.org/10.1016/j.enpol.2011.03.029

Singapore Building and Construction Authority. (n.d.). *Green Building Masterplan.* Retrieved June 17, 2022, from https://www1.bca.gov.sg/buildsg/sustainability/green-building-masterplans#:~:text=The%20SGBMP%20aims%20to%20deliver,80%2D80%20in%202030%E2%80%9D.&text=The%20earlier%20editions%20of%20the,Singapore's%20buildings%20have%20been%20greened

Singapore Building and Construction Authority. (2022). *Super Low Energy Programme.* https://www1.bca.gov.sg/buildsg/sustainability/super-low-energy-programme#:~:text=The%20SLE%20programme%20is%20the,other%20intelligent%20energy%20management%20strategies

Thomsen, T. U., Holmqvist, J., von Wallpach, S., Hemetsberger, A., & Belk, R. W. (2020). Conceptualizing unconventional luxury. *Journal of Business Research, 116*, 441–445.

Ulbrich, C. (2022, January 12). The conversation about green real estate is moving on as corporates prioritize sustainability. *World Economic Forum.* https://www.weforum.org/agenda/2022/01/green-real-estate-sustainability-corporate-priority/

United Nations Environment Programme. (n.d.). *Green financing.* Retrieved June 17, 2022, from https://www.unep.org/regions/asia-and-pacific/regional-initiatives/supporting-resource-efficiency/green-financing

United Nations Environment Programme. (2021). *2021 Global status report for buildings and construction-towards a zero-emissions, efficient and resilient buildings and construction sector.* United Nations Environment Programme.

Index

Note: **Bold** page numbers refer to tables; *italic* page numbers refer to figures.